S0-AXQ-066

THE FIFTH LANGUAGE

THE FIFTH LANGUAGE

Learning a Living in the Computer Age

ROBERT K. LOGAN

Published in 1995 by
Stoddart Publishing Co. Limited
34 Lesmill Road
Toronto, Canada
M3B 2T6
Tel. (416) 445-3333
Fax (416) 445-5967

Canadian Cataloguing in Publication Data

Logan, Robert K., 1939–
The fifth language: learning a living in the computer age

ISBN 0-7737-2907-0

1. Computers and civilization. 2. Information technology – Social aspects. 3. Information society. I. Title.

P96.T42L6 1995 303.48'33 C95-931152-1

Cover Design: Bill Douglas/The Bang
Computer Graphics: Tannice Goddard, S.O. Networking
Printed and bound in United States of America

Stoddart Publishing gratefully acknowledges the support of the Canada Council, the Ontario Ministry of Culture, Tourism, and Recreation, Ontario Arts Council, and Ontario Publishing Centre in the development of writing and publishing in Canada.

To my family,
the source of my inspiration,
with thanks to Margaret, Renée, David, Natalie,
Rebecca, Florence, and Nathan

CONTENTS

ACKNOWLEDGMENTS

I wish to give thanks, for their contribution to this project, to Angel Guerra and Don Bastian, who were the book's first champions. To my editor, David Kilgour, for his stimulating ideas and ruthless criticism. To my copy editor, Liba Berry, for her contribution to the clarity of the book. To my wife, Margaret, for her patience, for keeping me abreast of the latest developments in computing, and for giving me the chance to observe the computer-training business firsthand. To George Pòr, Bob Cooper, Manoo Missaghi, and Al Sokolow, who provided additional insights into the use of computing in business and were a source of fascinating perceptions and reality-based stories, and who contributed to the section on alignment in Chapter 8. And to Don Burrill, for his contribution to the section on computer literacy in Chapter 7.

THE FIFTH LANGUAGE

1

REDEFINING EDUCATION FOR THE COMPUTER AGE

> The future of work consists of learning a living in the automation age . . . Electric automation unites production, consumption and learning in an inextricable process . . . Paid learning is already becoming both the dominant employment and the source of new wealth in our society. (McLuhan 1964, 346–51)

Long before the advent of microcomputers, local area networks (LANs), and the Internet, Marshall McLuhan predicted that information technology would make dramatic changes in the structure of work and education. He was right. We are in the midst of a profound information revolution in which computer technology is changing the way we organize work and learn about our world. Knowledge, not capital, is the new source of wealth. Yet paradoxically, our education system is in total crisis. Our schools are not working because they do not provide students with the type of training and education they require to cope with today's economic realities. But this is only one aspect of the problem. Despite significant unemployment, jobs go unfilled because of the lack of skilled personnel, and many who are employed are not properly trained to do their jobs effectively. There is a significant mismatch between the needs of the workplace and what the educational system is delivering. The root of the problem is that we do not have a proper understanding of the role of education in the computer age. We are still operating on a model of education that is a holdover from the industrial age. Computing, telecommunications, and other forms of information technology have created a new environment and a whole new set of challenges.

The purpose of this book is to deal with the mismatches between school and learning, between education and work, and between computing and organizational structures inherited from the industrial age. It is about rethinking and reformulating the way we go about learning and preparing ourselves for work. It is a search for the new meaning of education. A cliché, perhaps, but a central question for a society in the midst of an information revolution.

In order to conduct the search for the meaning of education and its relevancy to work, we must examine the nature of cognition, or thinking, itself and how it is affected by language, information technology, communication media, socioeconomic institutions, and social-class structures. If the information environment in which we live is an extension of our psyches, as McLuhan claimed, and it is undergoing rapid and profound change, then so too is the way we think, process information, learn, organize and manage work, establish our values, set our goals, and formulate our aspirations. This book will examine all of these questions holistically. Not only do we have to question the relevancy of our education system, we must also examine our lifestyles. In the computer age, we can no longer compartmentalize our activities into education, work, and recreation as we did in the industrial era. This fragmentation no longer works, because of the speed at which technology changes plus the fact that work and the production of wealth is now knowledge-based. We can no longer divide work and learning. Life in the computer age only becomes meaningful when we integrate work, learning, and leisure time, whether we run a business, program a computer, or compose symphonies.

In describing the profound changes through which we are living, I have tried to strike a balance between optimism and pessimism. This is not a "technology gee whiz" book — the future is going to be automatically better, full of opportunities, and more satisfying. Nor is it a doom-and-gloom book, although for some caught in the avalanche of information, the future will be full of challenges. We face the possibility of electronic sweatshops, unemployment, and displacement. Computing and automation for all their efficiencies have increased the length of the work week, not reduced it. Many workers whose skills are overtaken by technology find themselves suddenly unemployed and, at a certain age, unemployable. This book tries to strike a balance between the service and disservice of information technology. It attempts to deal with reality as it is, not the hyped reality of the techno-enthusiasts nor the distorted reality of the techno-phobic and those displaced by technology. I hope the reader will better understand the current information revolution and as a consequence be able to take advantage of new opportunities rather than become a victim of the massive change which is wiping out many traditional jobs and ways of doing things.

This study began as an attempt to understand how computers could best be used in the educational arena. My research, conducted at the Ontario Institute for Studies in Education in Toronto (OISE), involved the classroom observation of the use of computers (Sullivan et al. 1986) and two theoretical studies (Logan et al. 1983; Logan 1986b) conducted for the Ontario Ministry of Education. My work in computer applications in education was carried out against the backdrop of work I had begun with Marshall McLuhan in which we studied the effects of the phonetic alphabet and literacy on the development of deductive logic, abstract science, codified law, and monotheism (McLuhan and Logan 1977). My approach to understanding computers in education was therefore strongly flavored by the communication theories of McLuhan. My own work in theoretical physics, on the other hand, which relied heavily on computers, made me equally sensitive to the information-processing aspect of computers. Therefore, at the very outset of my study to understand the impact of computers on education, I regarded the computer as both a communication medium and a tool for processing information.

At the same time that I was considering the role of computers in education, I was also extending the work I had begun with McLuhan on the alphabet to include the role that other written notations played in the history of ideas. I was fascinated by the work of Denise Schmandt-Besserat who showed that the origins of writing and mathematical notation can be traced to the small clay tokens that were used for accounting purposes in the Middle East from 8000 to 3000 B.C. (Schmandt-Besserat 1992). It was as a result of the consideration of her work along with my own on the alphabet effect (Logan 1986a) that I recognized that all written forms of language — whether it be writing or mathematical notation — have both a communications and an informatics dimension, as is also the case with computers. It was at this point that I formulated the two main theses of this book:

- all forms of verbal language have both a communications and an informatics dimension that facilitates human thought;
- speech, writing, mathematics, science, and computing form an evolutionary chain of languages.

In studying the history of communications and information processing, we will discover that each of these five forms of language has its own distinct semantics and syntax. Each evolved from its predecessors as new information-processing needs emerged that preceding languages could not deal with effectively. Each builds on the features of its predecessors while adding a number of new information-processing elements of its own. Each new language eventually led to an information explosion and a new set of challenges which set the stage for the next level of development and the emergence of still another form of language, culminating in the latest member of the set, computing, or the fifth language. By understanding the chain of events that led to the evolution of language and the emergence of the fifth language, we will be better able to understand computing's current impacts and predict some of its future effects.

I elected to examine the role of computing in education within the context of these two theses. As I began to probe my subject more deeply, I realized that the focus of my study was not only the role of computers in education but the much larger topic of the relationship between language and learning or communication and cognition.

My insight that computing contained elements of both communication and informatics seemed to apply to all forms of language. The term "informatics" is generally used only within the context of computers. It is defined in *Webster's*, however, as "the science of obtaining and transmitting information," and hence is not strictly limited to computers. One can therefore consider the informatics of the oral tradition or written texts as well as the informatics of mathematics and science.

These considerations led me to the realization that education should be basically concerned with the development of the skills associated with the use of all five modes of language — speech, writing, mathematics, science, and computing. I saw that the informatics side of language connected to cognition and that my study overlapped three basic domains: cognition, and hence informatics; communication, and hence language; and education, or learning.

My study of the role of computers in education had suddenly expanded in what seemed like a hundred different directions but at the same time converged on my previous studies of the impact of written notations on the development of ideas and social institutions. I had begun my study to find optimal ways to use computers in education and found myself using the insights that the use of computers provided to try to understand the nature of cognition, language, workplace organization, social class, and education. I began to feel as though I was losing the focus of my study when I discovered that part of my task was to tie together these different domains, which computing seems so naturally to integrate. I also realized that understanding the relationship between technology and cognition could provide insights that might help us devise a plan of action for education for both the school system and the workplace which would make them more responsive to the needs of an information-age economy.

The Four Facets of Computing

Computing is not just a new technology or a new medium of communication; rather, it is a radical new way to process and organize information and as such it represents a new form of language. To fully understand computing and its myriad impacts, we will study the computer from the perspectives of technology, communications, informatics, and language. In addition, we must also understand the relationships among these four aspects of computing. How do technology, communications, informatics, and language affect one another's development or evolution? To what extent are they distinct categories and to what extent are they just four different facets of human thought and organization?

THE SOCIAL DIMENSION

Still another dimension to computing is the socioeconomic one. The organization of work and social-class structure are influenced by the evolution of technology, informatics, and language. Agriculture led to new forms of labor and its organization, to new settlements and a two-class social system of landowners and serfs. With the rise of the city-state and the advent of numeracy and literacy, there arose a third class — the middle class. The middle class are not the middle-income earners but rather the class that adopted literacy and learning as a lifestyle and a way to earn a living. The bourgeois values of the middle class that are disdained by many artists and members of the intelligentsia are, in fact, the values that prize learning and literacy and that led to the scientific, philosophical, and artistic achievements that have enriched our civilization. It is also the middle class that led the Industrial Revolution and the current information revolution. Understanding the role of social-class structure is another element in developing our understanding of computing and its impact on education and work.

WHY OUR SCHOOLS DON'T WORK

Our schools are based on an industrial model with a delivery system patterned on the factory. Millions of schoolchildren are taught the same content in the same linear sequential order guided by a uniform curriculum dictated by a centralized bureaucracy at a municipal school board or state department (or provincial ministry) of education. Teachers continue to deliver an old style of book learning which does not take into account the nature of today's information-age economy or even some of the needs of day-to-day living. Students do not find enough relevancy in their schooling to take it seriously, which explains the high drop-out rate.

Not every aspect of the school system is a failure. Many students endure. Doctors, engineers, lawyers, and accountants continue to be trained. The success of our schools is limited, however. Today's educational system serves yesterday's needs by training managers and technicians to fit into a hierarchically structured industrial organization that is rapidly disappearing. Two practices of the schools are counterproductive: one is dividing the curriculum into subject disciplines and thereby encouraging specialism; the other is too much focus on content and not enough on process. Our school system must prepare individuals for the new demands of post-industrial society by providing them with both a knowledge base and a set of information-processing skills. It is unbelievable that most students graduate from university and/or high school with little or almost no computer skills at a time when computing is an indispensable part of life in the work place for the majority of professions. This, to me, is a major indictment of our education system.

Even though the failure of the school system has been recognized for a number of years, little, if any, remedial action to make structural changes has taken place. This is due in part to the conservative nature of the education community. The main reason for the failure, however, is that contemporary education lacks a mission that is in tune with today's social and economic environment, which has evolved and changed at a much faster rate than our schools. It is not so much that the content of the curriculum is out of date as it is that the style of education is not suited to contemporary needs and challenges. Stopgap measures like adding computer science to the curriculum or using computers as teaching tools will not remedy the ills of education. "Schools have already spent billions on computers — 97% of the 110,000 elementary and secondary schools in the U.S. now have them. But scores on standardized tests in math and reading are no higher than they were in 1970. The history of technology in education has not been a stellar one" (*Business Week*, Nov. 11, 1991, 158).

Approaches that consider how computers can be used to achieve present educational goals without considering how the entire educational enterprise should change are doomed to failure. The increased use of computers alone will not solve the problem. We must also learn to use computers differently than the way they are now being employed. "The evidence of our neglect of philosophy as it applies to educational computing lies in the absence of a significant body of discourse on what computers ought to be used for in education . . . Where is the literature debating what the aims or goals of computer education should be? Such literature exists, but it is extremely meager" (Maddux 1988, 6).

The role of the microcomputer in the school and the workplace needs to take on a new meaning. From the classroom perspective, the computer is not a tool to automate teaching but an environment in which students can learn and develop their cognitive skills through exploration and discovery. In the world of work, the computer is not just a tool to increase productivity but a learning environment in which employees continuously develop their information-processing and communication skills. Computers can facilitate a program of lifelong learning, another critical dimension in reformulating our thinking about education. Education must not stop once the student leaves school; it must become a lifelong process which integrates work and learning.

Integrating Work and Learning

If the key to survival in the computer age is to "learn a living" by integrating work and learning, then we must understand the relationship between education, work, and technology. Since knowledge workers and information technology are driving the changes in the information-age economy, the relationship between learning and work has to undergo radical change. Moreover, the rapid rate at which technology is changing is forcing the integration of education and work. Schools that provide students with knowledge of the latest developments in their field but do not equip them with the skills to update that knowledge are failing their students. On the other hand, employers who ignore the role of training and education in their operations are equally derelict in their duties to both their employees and their organization.

As the purpose of this book is to determine the most effective way to organize learning in our schools and in the workplace within the context of the new information environment of computers, we must reformulate the relationship between schooling, education, and work. Schooling and education are no longer synonymous. The model inherited from the industrial age within which school-based and industrial-based educators have traditionally operated is one in which learning and working are two separate activities that correspond to two distinct time periods in the life of an individual. In this model, the school years of childhood and early adulthood are when basic learning takes place and determines to a large extent what kind of work will be pursued and how far the student will go in life. With the exception of those lucky few who have participated in a co-op program, the work undertaken by students during their school years is usually unrelated to their professional development and is undertaken solely to earn money. The focus of working life, on the other hand, is not about learning but about earning a living. Some time during this period is devoted to training or professional development, depending on the nature of one's job and the attitude of one's employer. Because the focus of a business organization is on the bottom line, the amount of resources and employee time devoted to learning is frequently quite limited. The concept of "just-in-time training" in which an employee is only provided with enough instruction to perform the tasks at hand is a perfect example of the minimal attention paid to education by most employers.

The separation of the domain of schooling and childhood from that of work and adulthood can be traced to the early days of the industrial age. Before this period, formal education was provided for a tiny minority of the population who were the extremely privileged members of the aristocracy or the bourgeoisie. For the vast majority, there was no formal schooling and precious little childhood. Children began to work as soon as they could perform chores that would contribute to their families' survival. Their only learning was apprenticing with their parents to learn their trade.

It was the industrialization of Europe that led, beginning in England, to a system of formal education for the masses as well as to the invention of the concept of childhood. It was at this point that a clear separation was made between learning — that is, school learning — and work or holding a job. This separation was almost complete except for the short training period just after a student left school, during which time he learned the tricks of his trade. What he learned during this brief on-the-job apprenticeship would last a lifetime. This model has endured and still provides the foundation for our current education system and the induction of young people into the work force. But it is beginning to break down. The main reason is that the nature of work and the use of technology are changing. Perhaps more importantly, the rate of change is accelerating. Industrial-age schools were originally organized like factories and were mandated to mass-produce uniform and standardized graduates who could be easily integrated into the industrial system as both disciplined workers and consumers. The system worked because the methods and techniques of information processing which were paper-and-pen-based did not change within the time frame required to educate a young person. In fact, they did not change very much within the lifetime of an individual.

Now the rate of change in technology is forcing transitions in the education system. The idea of lifelong learning is slowly gaining credence. Unfortunately, however, it has not led to any structural changes in either the school system or the operation of the workplace. A great deal of lip service is paid to the idea, but neither the schools nor industry seem willing or able to commit the resources to make lifelong education a reality. Whatever commitment is now being made to lifelong education is being made by individuals. The long-term solution to this mismatch will require a serious commitment by both the school system and industry, but it would immediately solve two major problems: the irrelevancy of schooling as a preparation for work, and the need to constantly upgrade skills because of the rapid changeover of technology.

Eight Critical Questions

In order to achieve the goals of an integrated model of education and work based on the recognition of the unity of communications, cognition, and education, I have formulated eight basic questions that are critical for our study and will be addressed in this book:

1. How is computing changing work patterns and the organization of social institutions?
2. How can computers be used to achieve present educational goals?
3. How will present educational goals change to accommodate changes in the nature of work due to the widespread use of computers and other information technology?
4. How will computers change our notions of what we mean by communication, information processing, language, learning, education, and work?
5. How can the school system be restructured so that the goals of education and the vocational needs of society in the information age can be better matched?
6. How can computing be harnessed to inculcate students with the desire for lifelong learning?
7. How can work be organized to naturally promote learning?
8. How can education in the workplace be better organized to improve productivity so that learning becomes a lifelong activity and workers are properly trained to do their jobs?

Only by addressing these questions can we make our schools and workplaces relevant to our needs and create an atmosphere conducive to lifelong learning. Given that microcomputers have been in use in the workplace and in schools such a relatively short time, our understanding of this technology is still primitive. Educators and managers have only begun to explore the possible uses of microcomputers. This book, however, addresses the many issues raised by the eight questions above and provides a framework within which they may be addressed in the future so that computers may affect education and work in a positive way. The study begins by examining the ideas of Harold Innis and Marshall McLuhan who pioneered the study of the impact of communication media on social systems.

2

THE INNIS-McLUHAN COMMUNICATIONS REVOLUTION: The Method in the Madness of Marshall McLuhan

The Toronto School of Communications

The pioneering work of the Canadian communication theorists Harold Innis and Marshall McLuhan provide cogent insights into the cognitive and social impacts of communications media and other information-processing technologies on work and education. The general theoretical framework of this book, including the notion that a computer is both a medium of communication and an information-processing technology, grew out of the organic way in which Innis and McLuhan treated communications and their relationship to technology.

Innis and McLuhan have also influenced a number of other fields including commerce, politics, advertising, social sciences, and the arts. Their work was more concerned with the larger patterns of communication rather than the meticulous analysis of detailed data. They made use of intuition and poetic insight and were given to speculation and generalization. Some have suggested that they were unsystematic in their approach and that they lacked a methodology. Although they did not articulate and separate their methodology and assumptions from their analyses, they nevertheless developed a systematic approach to media studies and communication theory which they employed consistently and coherently.

This chapter has four objectives. The first is to show that the work of Innis and McLuhan gave rise to a coherent school of thought, the Toronto School of Communications, with a systematic methodology and a common set of basic assumptions and presuppositions. The second is to articulate these assumptions, presuppositions, and methods based on the writings of Innis and McLuhan. The third is to update their approach to communications by extending their methodology to include the most recent breakthroughs in communications and informatics due to microelectronics. Finally, the fourth objective is to develop a general framework for and specific insights into the way in which the workplace and the education system can be reformed to meet the needs of the information age, and, it is hoped, to evolve into new models of work and education.

The Assumptions and Methods of the Toronto School

Harold Innis and Marshall McLuhan were, respectively, professors in the political economy and English departments at the University of Toronto, where they met, interacted, and influenced each other's work. McLuhan freely acknowledged the impact that Innis had on him. In the foreword to the 1972 edition of Innis's *The Bias of Communication*, McLuhan wrote: "I am pleased to think of my own book *The Gutenberg Galaxy* as a footnote to the observations of Innis on the subject of the psychic and social consequences, first of writing and then of printing. Flattered by the attention that Innis had directed to some work of mine, I turned for the first time to his work. It was my good fortune to begin with *Minerva's Owl*. How exciting it was to encounter a writer whose every phrase invited prolonged meditation and exploration" (Innis 1972, ix, McLuhan's foreword).

Perhaps this quote explains why the work of Innis and McLuhan forms a coherent whole, exhibiting a unified purpose, set of assumptions, and methodology. Their work has inspired other scholars, many working in Toronto, to pursue similar lines of research in this tradition, and hence, the designation, the Toronto School of Communications. That these two giants of communication both came from Canada may be a coincidence or it may reflect the fact that, with the world's second-largest land mass and a relatively sparse population, Canada has always been challenged by the problems of transportation and communication.

In articulating the assumptions and methods of Innis and McLuhan, I have not attempted explicitly to separate the elements of their methodology from their basic assumptions. Innis and McLuhan developed their methodology intuitively. Rather than deducing their methods from a basic set of axioms, they arrived at their methodology inductively on the basis of their respective observations and experiences as an economic historian and a literary critic.

Innis arrived at his methodology through his concern with trade and commercial patterns in the economic world. McLuhan developed a parallel methodology by studying how writers affect their reading public. The methods and assumptions of their two independent approaches, however, display an enormous number of parallels and contain a significant number of overlapping claims and conclusions. Innis turned to the study of communication per se towards the end of his career, whereas McLuhan pursued media studies almost from the outset of his academic life.

The order in which I have described their thoughts reflects my notion of a systematic presentation of their ideas. The ideas are presented sequentially, since this form of ordering lends itself to the written word. I can assure the reader from my own experience of working with McLuhan, however, that he never analyzed a problem sequentially and certainly not in the order presented here. All of the methods set out below were brought to bear simultaneously on whatever issue or challenge he was tackling.

MEDIA AS LIVING VORTICES OF POWER

The unique element in the Innis–McLuhan approach to media studies, and hence the logical starting point for articulating their methodology, is their attitude towards the dynamic role that media and technology play in economic, political, social, and cultural environments. Rather than viewing media as passive conduits for carrying information or communicating ideas, the two theorists regarded media, as McLuhan put it, as "living vortices of power creating hidden environments (and effects) that act abrasively and destructively on older forms of culture" (Innis 1972, v, McLuhan's foreword). This applies to computing and other information-processing technologies; thus the need to reevaluate our workplace and education system. The older form of culture which the school system was intended to serve has been transformed by the new technologies, and a new model for education and its relation to work must be found.

MEDIA CREATE NEW SOCIAL PATTERNS AND RESTRUCTURE PERCEPTIONS

As "living vortices of power," media create new social patterns. Innis wrote that "a medium of communication has an important influence on the dissemination of knowledge over space and over time and it becomes necessary to study its characteristics in order to appraise its influence in its cultural setting" (Innis 1951, 33). Once the dominant media and technologies of a culture are known, one knows "the cause and shaping force of the entire structure

[and] what the pattern of any culture [has] to be, both psychically and socially" (Innis 1972, xii, McLuhan's foreword).

Media not only have an impact on social patterns but also directly affect the psyche and the ways in which people think and learn. "The effects of technology do not occur at the level of opinions or concepts, but alter sense ratios or patterns of perception steadily and without any resistance" (McLuhan 1964, 18). A medium of communication creates a sensory bias, and hence a new cognitive style. Preliterate cultures existed in an acoustical world where all information is processed simultaneously in real time. Literate societies, on the other hand, developed a visual bias in their perceptions because literacy involves the use of the visual faculty in which information is processed in a linear sequential manner letter by letter or word by word. With the electronic dissemination of information, new sense ratios will emerge in which both the visual and the acoustical will come into play; the visual because of the continued use of abstract signs — the letters of the alphabet and/or numerals — and the acoustical because of the real-time simultaneity that instantaneous electronic communication makes possible.

"THE MEDIUM IS THE MESSAGE"

Like most good aphorisms, McLuhan's famous dictum has more than one meaning. Two of its principal interpretations form an integral part of the philosophy of the Toronto School of Communication. One is the notion that, independent of its content or messages, a medium has its own intrinsic effects on our perceptions which are its unique message. "The message of any medium or technology is the change of scale or pace or pattern that it introduces into human affairs." McLuhan cites the way the railway created "totally new kinds of cities and new kinds of work" (McLuhan 1964, 8). What McLuhan writes about the railroad applies with equal force to the media of print, television, and the microcomputer. "The medium is the message" because it is the "medium that shapes and controls the scale and form of human association and action" (McLuhan 1964, 9). The effects of a medium impose a new environment and set of sensibilities upon its users. For example, microcomputers in the classroom and the workplace affect social interactions because the display monitors encourage and facilitate sharing.

"The medium is the message" also carries the notion that a medium transforms its message or content. A movie shown on television or a play that is filmed affects its audience differently from the original. The classroom drill exercise in a workbook loses some of its tedium when it is programmed onto a microcomputer.

"MEDIA ANALYSIS" VERSUS "CONTENT ANALYSIS"

The aphorism "the medium is the message" has been subject to much misinterpretation, particularly by those who take it to mean that the content of a medium is to be totally disregarded and the only message is that of the medium itself. But McLuhan writes: "The latest approach to media study considers not only the 'content' but the medium and the cultural matrix within which the particular medium operates" (McLuhan 1964, 11). This clearly indicates that McLuhan's intention was not to wholly ignore the content of a medium in analyzing its impacts. McLuhan, who enjoyed both hyperbole and paradox, was anxious, however, to shift the focus of analysis from that solely of content to one that also included the effects of a medium independent of its content. He believed that the study of communication patterns requires a balance between "media analysis" and "content analysis." The first step in using a new medium in an educational setting is to learn as much as possible about the medium itself. Once the medium and the skills required to use it are understood, one can then use the medium to acquire new knowledge. The actual content of a new computer-configured education system and workplace will not change that dramatically from the content of the past, but the style of delivery and the approach to that content will be radically different.

THE CONTENT OF ANY MEDIUM IS ALWAYS ANOTHER MEDIUM

The content of the telegraph is print, and the content of print is writing. The content of writing is speech, and the content of speech is thought (McLuhan 1964, 8). When a medium first appears, it uses the content of another medium exclusively for its content until its users have learned to exploit the new medium to develop new forms of expression. Innis and McLuhan provide a number of historical examples to illustrate that the immediate content of a new medium is the content of an older one: "The narrow two- to three-inch column of the papyrus roll was continued in the early vellum manuscripts, which contained three or four columns to a page" (Innis 1951, 116). The press was first used to print ancient and medieval books rather than produce new books. Printed books were often taken to a scribe to be copied and illustrated (Steinberg 1955).

The first literature to appear in writing were transcriptions of material that had been composed orally, such as Homer's *Iliad*, the Hebrew Scriptures, and the Hindu Vedas. Only later did authors develop a prose style composed expressly for the written medium. A similar pattern developed with television. Movies and vaudeville were the first content of TV; only later were shows like Rowan and Martin's *Laugh In* or *Second City TV* (SCTV) crafted specifically to take advantage of the medium. At first, the content of computers was print, mathematics, and scientific data. In the commercial world, computers were first used to perform complex calculations and manipulations of data. It is only in recent years that applications specifically designed for computers, like spreadsheets and relational databases, have emerged. Microcomputers were first employed in the classroom as electronic flash cards for drill and practice. It was only later that new educational applications such as word processing, simulations, and hypertext came to be used.

THE REVERSAL OF CAUSE AND EFFECT

In the Innis–McLuhan approach, a medium or technology is studied by examining its effects not only on its direct users but on a society as a whole. One of the ways of understanding the effects of a technology is to consider the changes that would take place if that technology suddenly disappeared. With Innis and McLuhan, the notion of a one-to-one correspondence between cause and effect gives way to a "total-field-theory" approach where one and the same cause may have a wide diversity of predictable and characteristic effects (Innis 1972, ix, McLuhan's foreword). Rather than "matching one cause against another effect [Innis and McLuhan] went on making sense of the process released by the new structures of communication" (Innis 1972, x, McLuhan's foreword). Linear sequential cause and effect, brought into doubt by Hume's skepticism and discounted as a principle of causality by quantum mechanics, is replaced by an interplay of cause and effects. "Instead of asking which came first, the chicken or the egg, it suddenly seemed that a chicken was an egg's idea for getting more eggs" (McLuhan 1964, 12). A similar idea was also developed in Richard Dawkins's *The Selfish Gene.* The notion that the gene is a mechanism to propagate the organism was reversed and evidence was presented to support the proposition that the organism is merely a mechanism for a gene to propagate more of its own kind (Dawkins 1976).

The reversal of cause and effect is the technique of the artist who begins with the effect he or she wants to create in his or her audience and invents a work of art that achieves that end. The reversal of cause and effect is one of the consequences of electronically configured information patterns and the need for the cybernetic feedback and feedforward of information. The speedup of information flow requires planners to have complete knowledge of all possible ultimate effects, and hence the need for the reversal of cause and effect. This reversal also led to the nineteenth-century discovery of the technique of discovery, that is, starting with the thing to be discovered and working backwards to reproduce the desired object. Reverse engineering, the process whereby a competing firm takes a finished product and figures out how to manufacture it, is yet another example of the application of this technique.

The reversal of cause and effect is the only way that we can reform the school system and develop a new model of education. We must work backwards from the desired effects if we wish to find the curriculum that will serve the needs of the information age.

FIGURE-GROUND

The meaning of any "figure," whether it is a technology, an institution, a communication event, a text, or a body of ideas, cannot be determined if that figure is considered in isolation from the ground or environment in which it operates. The ground provides the context from which the full meaning or significance of a figure emerges. The concern with the relation of figure to ground explains the emphasis of the Toronto School on interface and pattern rather than on a fixed point of view.

The following examples illustrate the way in which the context can transform the meaning of a figure: a smokestack belching smoke, once a symbol of industrial progress, is today a symbol of pollution; the steam locomotive, once the symbol of mechanical power, is now an art object and a focus of nostalgia; the suntan, once a sign of hard work in the fields, is now a symbol of affluence and holidaying and will probably evolve into a symbol of reckless disregard for health concerns such as skin cancer.

The figure of the school, which has been slow to change, has taken on a new meaning because of the radical changes in the information environment in which it operates. The school system, once considered an engine of social mobility and a fount of opportunity, is now regarded by many students as an obstacle or a prison in which one must do time until one is released into the real world as an adult. In order to succeed, the figures of the school and workplace must change so that they are back in synch with the real world, a world that is undergoing continuous and rapid change due to the accelerated flow of information.

INTERFACE AND PATTERN RATHER THAN A "POINT OF VIEW"

A distinguishing characteristic of the Toronto School is the absence of the traditional methodology of orthodox scholarship whereby a particular "point of view" or hypothesis is articulated and all subsequent observations are made to support or refute it. McLuhan describes his approach of working without a "point of view" as "observation minus ideas" (McLuhan 1977). A point of view ties one down to a single perspective and limits the richness of insight that one can develop. A point of view represents the visual bias of the written word rather than the simultaneous, all-at-once, acoustic flow of information characteristic of both the electric information age and the oral tradition. A point of view is equivalent to the "single vision of Newton."

McLuhan described how Innis articulated the alternative stance to adopting a point of view. "He changed his procedure from working with a 'point of view' to that of the generating of insights by the method of 'interface.' By contrast, a 'point of view' is merely a way of looking at something. But an insight is the sudden awareness of a complex process of interaction" (Innis 1972, viii, McLuhan's foreword). The point-of-view approach leads to classification, a left-brain process, whereas working with interface leads to pattern recognition, a right-brain process.

The point-of-view approach corresponds to what Thomas Kuhn calls "normal science" (Kuhn 1972). Once a revolutionary idea succeeds in describing a new set of data or phenomena, a new paradigm develops which is imitated and articulated in as many areas of application as possible. The development of a new paradigm is "revolutionary science" and requires the lack of a point of view. "Normal science," or the articulation of the new paradigm, involves the adoption of the new paradigm as a point of view. The Toronto School attempts to operate in the revolutionary mode without a point of view.

The school system, on the other hand, operates by transmitting the paradigms of the past. It teaches by example and encourages the adoption of the point of view of others. This approach worked well during the industrial age when ideas were slow in changing. In order to accommodate the rapid change characteristic of our times, we must train our young people to be innovative and open to new ideas, and not to rely necessarily on past points of view. The same applies to business, which also operates on the paradigms of the past but which must change to accommodate the new reality of microcomputing and connectivity.

THE SERVICE AND DISSERVICE OF NEW MEDIA

New media create both service and disservice. This includes the institution of schools, the corporation, and the medium of the computer. The new services of a medium are obvious since the purpose of any invention is to enhance or augment some activity of humankind. Each medium, however, is a "vortex of power" creating new environments and acting abrasively on older cultural forms. The potential for disservice and the destruction of cultural forms is enormous. One need only consider the effect of the speedup of information on the school system to see the destructive power of a new technology.

Consider the service and disservice of logic. Logic helped create a new scientific and rational way of thinking that led to many advances. However, the rational thought of the Greek philosophers, based on the technologies of the phonetic alphabet and deductive logic, also acted abrasively and destructively on the oral tradition. Indeed, Plato would have banned poets from his utopian Republic. The disservice of the rational approach did not extend merely to the destruction of the older order of the oral tradition.

The excesses of being overly logical also impeded developments within the new order of thinking. Parmenides' arguments against the possibility of non-being on logical grounds inhibited the invention of zero by the ancient Greeks. The overly abstract approach of the Greek philosophers prevented the development of an empirical approach to science and likely delayed the scientific revolution. These examples illustrate that too much of a good thing can be a disservice.

THE ABSENCE OF A MORAL JUDGMENT

The Toronto School attempts to articulate the patterns of change that occur as new media create innovation and push out older forms without expressing a particular point of view or making a moral judgment. McLuhan held that "moral and emotional indignation was simply an indulgence on the part of those powerless to either act or to understand" (Marchand 1989, 121).

Because he described the power of electric technology and the threat that it posed to literacy, McLuhan was often accused of being an enemy of the book and an advocate of electronic media, a somewhat ironic epithet for one of the most widely read scholars of our time. McLuhan, in fact, was extremely critical of television and felt its use by children should be carefully regulated by their parents. He was not passing judgment on the transition into the electric age of information with its potential threat to literacy; rather, he was warning society of the powerful changes that electric forms of information were creating. Like the messenger of bad news who was killed by the king, he was attacked as though he were somehow responsible for encouraging the new phenomena. The same treatment is often directed at critics of the present-day school system.

MEDIA STUDIES AS CIVIL DEFENSE AGAINST MEDIA FALLOUT

"Is not the essence of education civil defense against media fallout?" (McLuhan 1962, 294 [Mentor ed.]). "What Innis indicates as a basis for social survival is nothing less than a reorganization of our perceptual lives and a recognition that the environments we witlessly or involuntarily create by our innovations are both services and disservices that make very heavy demands of our awareness and understanding" (Innis 1972, vii, McLuhan's foreword). It is only by studying various media and their effects that one can anticipate change and be prepared for it. "Control over change would seem to consist in moving not with it but ahead of it" (McLuhan 1964, 199). The school system and to some extent business, because of the intrinsically conservative nature of both institutions, tend to move with change rather than ahead of it.

THE SUBLIMINAL EFFECTS OF MEDIA

The fish is unaware of the water in which it swims. It is blind to its environment — its medium. We, too, function and perceive reality in an information/communication environment which we take for granted. People do not realize that they live in a small village until they have been to a big city (Hall 1973). The fish only appreciates its medium once it has been forced out of the water.

Innis and McLuhan were aware of the "nemesis of creativity" which blinds a culture to its most significant inventions. "The Greek dramatists presented the idea of creativity as creating, also, its own kind of blindness, as in the case of Oedipus Rex, who solved the riddle of the Sphinx" (McLuhan 1964, 39). "It is one of the ironies of Western man that he has never felt any concern about inventions as a threat to his way of life. The fact is that, from the alphabet to the motorcar, Western man has been steadily refashioned in a slow technological explosion that was extended over 2500 years" (McLuhan 1964, 270). This time around we are being refashioned by a rapid explosion of information.

McLuhan describes the insensitivity to new media and technology as numbness: "The electric technology is within the gates, and we are numb, deaf, blind, and mute about its encounter with the Gutenberg technology" (McLuhan 1964, 17). This numbness he attributes to the need to protect the psyche when it is suddenly exposed to the new "stress of pace" which a new medium creates. This numbness in the case of electric technology is described as a self-amputation of the nervous system which becomes extended by the new medium. It is because of this self-induced numbness that the effects of media are often subliminal and hidden. Teachers continue to conduct their classes as though we were still in the midst of the industrial age. Businesses, even those in the information sector like IBM, conduct themselves along the patterns of the nineteenth century.

THE COUNTERINTUITIVE EFFECT OF MEDIA

The effects of media are complex because of the vast energies they release, the wide scope of their impact, and their subliminal natures. It is therefore understandable that, frequently, their impacts are counterintuitive. Henry Ford thought that by making the Model T affordable, Americans would be able to enjoy the countryside; instead, the automobile destroyed much of the countryside. The counterintuitive effects of technology occur because their side effects overwhelm their intended or desired function. The increased access to information made possible by computers has flipped into information overload because the approach to the use of the computer is based on an industrial-age mentality and attitude.

LAWS OF THE MEDIA

McLuhan developed a set of rules which he called the Laws of the Media (LOM) (McLuhan and McLuhan 1988) for studying the effects of media or technologies which specifically illustrate their counterintuitive nature. LOM consists of the following four laws:

1. Every medium or technology enhances some human function.
2. In doing so, it obsolesces some former medium or

technology which was used to achieve the function earlier.

3. In achieving its function, the new medium or technology retrieves some older form from the past.
4. When pushed far enough, the new medium or technology flips or reverses into a complementary form.

To illustrate a LOM, let us first consider the medium of money, which enhanced trade and commerce, obsolesced the barter system, retrieved the conspicuous consumption of hunting and gathering societies, and flipped into credit.

Or consider the technology of the automobile, which enhanced transportation, obsolesced the horse and carriage, retrieved the knight in shining armor, and flipped into the traffic jam — the lack of mobility.

The laws apply with equal validity to all the artifacts of humankind, whether they are communication media, technological inventions, or scientific laws or principles. The term "obsolescence" used in this context requires some clarification because the word is often misinterpreted or interpreted too literally. The obsolescence of the barter system by money does not mean that the use of cash forever ended the straight exchange of commodities. It does mean, however, that cash transactions became the dominant mode of commercial exchange. Similarly, the obsolescence of the horse and carriage did not end forever this mode of transportation, which is still used as a tourist attraction in large cities and as a mode of transport in many third-world centers or among certain cultural groups. The automobile, however, has come to dominate traffic patterns.

This point becomes particularly important when considering the obsolescence of print by electric information. When McLuhan indicated that electricity had obsolesced print, defenders of the book pointed to statistics that showed an increase in the sale of books and the number of new titles available. They argued that the book had not been obsolesced but was alive and well. What these critics failed to grasp is that electric information has come to dominate print as the principal mode of distributing information and that information consumers spend far more time with electric media such as radio and television than they do with books. They also failed to take into account that the production of print material was also being more and more influenced by the electronic telecommunication, handling, and storage of information. When writing obsolesced the oral tradition, people did not suddenly stop talking to each other. But writing came to dominate the spoken word whenever a permanent record of information was desired or communication over distance was required.

A technology or medium does not disappear when it becomes obsolesced; it often becomes an art form or a source of nostalgia. The phenomena of antique reproductions and hand-crafted utensils as art forms illustrates this idea. And in the age of supermarkets and slick packaging, a farmer's market becomes a source of intense nostalgia and a tourist attraction.

How do the laws of the media apply to computers? Computers enhance the individual's access to information, "retrieve" the generalist's approach to processing information, and hence obsolesce the specialist or the expert. Computers retrieve individual learning and hence obsolesce mass education. The computer, if pushed far enough, will flip into information overload in which the quantity of information overwhelms the user and no pattern becomes discernible. The challenge facing schools and the workplace, therefore, is to provide context and relevance to information and "learning."

THE EQUIVALENCE OF MEDIA AND TECHNOLOGIES

As the Laws of the Media apply with equal validity to media and to tools, the distinction between technological inventions and media of communication is somewhat arbitrary. I use "technology" in its broadest sense to include not only hardware (machinery) but also all forms of communication and information processing, including speech, writing, mathematics, and science. The term stems from the ancient Greek word *technologia*, which means a systematic treatment.

Media such as the book, the telephone, radio, and television differ from tools such as the hammer, the bulldozer, the airplane, and the light bulb, but there are also some very important overlaps. The most obvious is that all media function as tools serving our needs and all consist of some form of technology. One can also argue, however, that technologies become media, for example in the case of the light bulb when it is used to spell out advertising slogans. The road, the canal, and the railroad are also technologies that serve as media (grounds) for the automobile, the ship, and the train, respectively. The automobile, the ship, and the train have as their content passengers and freight. Moreover, the automobile functions as another kind of medium when it becomes a status symbol, a symbol of teenage rebellion, a symbol of a macho man's potency, or a haven for privacy. These meanings or functions of the automobile are often as important as its primary "message": mobility.

TECHNOLOGY AS EXTENSIONS OF THE BODY AND MEDIA AS EXTENSIONS OF THE PSYCHE

From the Laws of the Media we discovered that technologies and media enhance human functions. Mechanical technologies serve to extend our physical bodies, while media of communication extend our psyches or nervous systems. So computing may be considered as a form of "artificial intelligence" (AI). The confusion with the term "AI" arises when it is used to denote an independent intelligence separate from its human creators and users. Those who indulge in this futuristic fantasy forget that computers are only extensions of our own intelligence.

McLuhan observed that mechanical technologies extended our bodies in space and that electric technolog extended our central nervous system into what he termed "a global embrace, abolishing both space and time" (McLuhan 1964, 3). The integrated nature of the total communications/technology environment is a consequence of this fact. "Since all media are extensions of ourselves, or translations of some part of us into various materials, any study of one medium helps us to understand all others" (McLuhan 1964, 139). Media therefore demand a holistic and multidisciplinary treatment if their impact on human interactions is to be properly understood. When considering the impact on education, we must consider how computers will change the nature of literature, mathematics, and science. When considering the impact on business, we must consider how computers will integrate and align the processes of product design, manufacturing, distribution, marketing, and customer support.

THE FLIP: HUMANKIND AS AN EXTENSION OF ITS TECHNOLOGIES

"To behold, use or perceive any extension of ourselves in technological forms is necessarily to embrace it. By continuously embracing technologies, we relate ourselves to them as servo-mechanisms" (McLuhan 1964, 46). At first, technology serves as an extension of humankind and then suddenly a flip occurs and humankind is transformed into an extension of its technology. In terms of the Laws of the Media, the flip occurs in the following sequence: technology enhances some function or extension of man, obsolesces some part of man's body, retrieves the gods in the sense that it provides us with the power to control nature, and, pushed far enough, flips into "machine in control of man" or "man as an extension of his technology."

To illustrate this point, consider whether alien beings monitoring activities on this planet would be able to immediately determine who was in control of the planet: cars or people. Indeed, it is difficult to determine whether people serve the automobile or the automobile serves people. Automobiles and people seem to receive equal attention. In terms of energy expenditures, the car, in fact, receives more kilocalories than people. Modern industrial man spends 25 percent of his time providing himself with transportation compared with only 5 percent for members of hunting and gathering societies. The car has taken over modern life in terms of how we spend our time and how our urban and rural landscapes have been transformed by our need for mobility. We have become extensions of our cars, their servo-mechanisms. The same argument can be made about computers. Many of our institutions will have to change in order to accommodate the use of computer technology. We have no alternative, unless we wish to forgo the advantages that computers provide. But we must use computers wisely, recognizing what we give up when we submit to their tyranny.

HISTORY AS THE LABORATORY OF MEDIA STUDIES

History is used as a dynamic tool in the study of media by the Toronto School. McLuhan wrote: "[Innis] had discovered a means of using historical situations as a lab in which to test the character of technology in the shaping of cultures. Innis taught us how to use the bias of culture and communication as an instrument of research. By directing attention to the bias or distorting power of the dominant imagery and technology of any culture, he showed us how to understand cultures" (Innis 1972, xi, McLuhan's foreword). McLuhan felt that an understanding of history was essential for understanding the future and the impact of new technologies. He often used the metaphor of the rearview mirror, a device by which we are able to determine what is about to overtake us from our past. Furthermore, according to McLuhan, history is not to be regarded as a series of events but rather as a dynamic process with a discernible pattern which repeats itself from culture to culture and from technology to technology.

MEDIA AS STAPLES

In *Empire and Communications*, Innis described socioeconomic and political systems in the context of their dominant forms of communication. He explained the difference between the imperialistic military bureaucracy of Rome and the conservative priestly bureaucracy of Babylon in terms of the materials upon which they wrote. The Babylonians used clay tablets, which provided a permanent written record of their culture and hence command over time. The Romans, on the other hand, wrote on an extremely portable medium, papyrus, which gave them command over space.

In their historical analyses, Innis and McLuhan regard technological media as staples or natural resources, exactly like wheat, cotton, or oil. Just as societies came into conflict over staples (wheat versus cattle in the American West) so, too, can conflicting media become a source of confrontation. The conflicts of the eighteenth and nineteenth centuries between oral cultures and industrialized society have given way in the twentieth century to the clash of mechanical or industrial forms of organization with electronic or postindustrial modes of operation in both the workplace and the school system.

THE THREE COMMUNICATION AGES OF HUMANKIND

McLuhan, building on the ideas of Innis, divided human history into three distinct periods based on the modes of sensibilities their media made available to them. The first era, that of the oral tradition, stretches from the time humankind first acquired speech to the beginnings of literacy five thousand years ago. The second era, the age of literacy, includes the period from the invention of writing to the discovery of electricity and its use in the form of the telegraph. The age of writing is further subdivided into three periods, the first beginning with the advent of written symbols, the second with the invention of the phonetic alphabet, and the last with the invention of the printing press. The third communication era, that of the electric flow of information, covers the period from the first use of the telegraph in 1844 to the present. "We live today in the Age of Information and Communication because electric media instantly and constantly create a total field of interacting events in which all men participate" (McLuhan 1964, 248).

Innis also divided history into periods in which different modes of communication dominated. He divided the age of literacy, however, by the nature of the medium upon which texts were written. "We can conveniently divide the history of the West into the writing and the printing periods. In the writing period we can note the importance of various media such as the clay tablet of Mesopotamia, the papyrus roll in the Egyptian and in the Graeco-Roman world, parchment codex in the late Graeco-Roman world and the early Middle Ages, and paper after its introduction in the Western world from China" (Innis 1972, 7).

During each of these three major communication eras, socio-economic and cultural life were deeply affected by the dominant medium of communication. As each new mode of communication had an impact on society, it dominated those which had preceded it, not obsolescing them, but dramatically changing their character and the use to which they were put. Speech or the oral tradition naturally survived both literacy and electricity but its function changed. It retained its dominance for conversation and everyday communication; however, it was no longer used as the repository of a culture's traditions as it had been in Homer's time, nor the means of spreading news from one village or country to another. The wandering minstrel who had conveyed information across both space and time was replaced by the written record which spanned space through the courier and time through the library or archive. With writing, the spoken word took on new functions, sometimes becoming an art form in the guise of poetry and theatre.

Writing also underwent enormous changes with the advent of electricity. The modern newspaper, as McLuhan points out, is a product of the printing press and the telegraph. Electricity and the instantaneous flow of information changed the psychic environment of authors, causing them, in McLuhan's words, to "live mythically and in depth" (McLuhan 1964, vii). As a consequence, writers became concerned with psychology, anthropology, and sociology. The psychological novel and stream-of-consciousness technique were born. Electricity produced another unexpected flip that affected the spoken word, namely a revival of the oral tradition in the art, music, and literary world of the nineteenth and early twentieth centuries as exemplified by jazz and the use of African masks in Cubism.

The structure of education also changed with the advent of each new mode of communication. With writing, education was transformed from the apprenticeship mode of learning on the job to formal schooling. The alphabet, abstract science, formal logic, and codified law gave rise to the academies of higher learning that appeared in ancient Babylon, Egypt, and Greece. The printing press gave rise to the modern school system of mass education and the assembly-line style of mass production. Electronic media and computers will also inspire major innovations, the nature of which is still to be determined.

WRITING, THE ALPHABET, AND THE PRINTING PRESS

McLuhan attributed the phonetic alphabet's power to its subliminal effects. Of all the writing systems, the phonetic alphabet permits the most economical transcription of speech into a written code. It also provides its users with the model for abstraction, analysis, coding, and decoding which came to characterize Western thought. The phonetic alphabet introduced a double level of abstraction in writing. Words are divided into the meaningless phonemic (sound) elements of which they are composed and then these meaningless phonemic elements are represented visually with equally meaningless signs, namely, the letters of the alphabet. This encourages abstraction, analysis (since each word is broken down into its basic phonemes), coding (since spoken words are coded by visual signs), and decoding (since those visual signs are transformed back to spoken sounds through reading).

In an article entitled "Alphabet, Mother of Invention," McLuhan and I indicated some of the cultural impacts that resulted from the invention of phonetic writing (McLuhan and Logan 1977). The Mesopotamian phonetic syllabary inspired the organization of social mores into forms of codified law, the most famous of which is the Hammurabic code. The impact on the Hebrews of the alphabet which they borrowed from the Midianites was immediate and dramatic. In addition to bringing writing to the Hebrew children, Moses also brought them codified law, or the Ten Commandments, as well as a more abstract and monotheistic concept of God. These three developments occurred at the same time, as symbolized in the Bible by the story of Moses receiving the Ten Commandments on Mount Sinai: "He gave Moses, when He had made an end to communing with him upon Mount Sinai, two tables of testimony, two tables of stone, written with the finger of God" (Ex. 31:18).

The introduction of the phonetic alphabet into ancient Greek society had an equally dramatic effect on that culture. Within 500 years of the transmission of the alphabet from the Phoenicians, the Greeks developed the main intellectual concepts which have formed the foundation of Western civilization. They created, for the first time in history, abstract science, formal logic, axiomatic geometry, rational philosophy, representational art, and individualism. While not suggesting a causal connection between these developments and the alphabet, I believe that the alphabet, by serving as a paradigm for classification, analysis, and codification, created the conditions that made these new ideas possible; it also explains why abstract science began in the West and not the East, despite the superior achievements of the Chinese in technical matters, such as their invention of silk, paper, gunpowder, metallurgy, porcelain, animal harnesses, and irrigation systems, to mention a few.

The most dramatic communication revolution to follow the introduction of the alphabet was the Gutenberg printing press. The revolutionary changes brought about by this technology have been documented by McLuhan in *The Gutenberg Galaxy*, a seminal work in which he shows the impact of print on such major cultural transformations as the rise of science, the Reformation, the Enlightenment, the rise of nationalism, and the Industrial Revolution. "The invention of typography confirmed and extended the new visual stress of applied knowledge providing the first uniformly repeatable 'commodity,' the first assembly-line, and the first mass production" (McLuhan 1962, 153).

FIVE TECHNOLOGICAL BREAKTHROUGHS

The effect of electricity is to speed up the process that it enhances, whether this be communication at a distance via telegraph, information storage via magnetic tapes, or information feedback via the thermostat. The impact of the speedup of information on each of the activities it is affecting is not always the same, however. McLuhan did not explicitly distinguish between the various phases of the electric information age; rather he tended to treat all electric forms of information uniformly. He lumped together a number of separate and distinct breakthroughs perhaps because each involved the electric speedup of information. "As technology changed, McLuhan did not appear interested in exploring the implications of the latest advances — not even the great technological wonder of the seventies, the microchip computer. For him, electronic technology was largely broadcasting — instantaneous, global communications" (Marchand 1989, 253).

Perhaps it was because of the integrative nature of electricity that McLuhan failed to distinguish the various technological breakthroughs associated with the electric information age. Greater insights into each of these individual innovations can be achieved, however, by examining them separately, in much the same way that making a distinction between ideographic and phonetic writing or manuscript literacy and print also proved useful. A better understanding of the nature of contemporary information and communication systems and their impacts can be obtained, therefore, by distinguishing five separate technological breakthroughs (Logan 1982) that can be associated with the electrification of information. These are: (1) telecommunication; (2) information storage; (3) automation or cybernetics; (4) the general systems approach; and (5) microelectronics. This analysis represents an extension of the ideas of Innis and McLuhan to include computer technology within the framework of their approach to communications (Logan 1982).

The first electrification of information came in the form of telecommunications as represented by the telegraph. This was followed by the telephone, the microphone and loudspeaker, the radio, and finally television. The second breakthrough occurred with the electrification of information storage first in the form of the phonograph record, then in the form of magnetic tape. The next breakthrough, automation, occurred with the computer and other cybernetic control devices providing the electric feedback of information. The fourth element is the operations research or general systems approach in organizing human activities or projects and includes the emergence of software on a par with hardware. The fifth breakthrough, microelectronics, entails the miniaturization of the computer and a merger of hardware and software into a single integrated circuit imprinted on a silicon chip. Having identified the five breakthroughs which created the computer revolution, we will now examine each one in detail.

1. Telecommunications: Each of the communication and information breakthroughs brought about by the application of electricity has its roots in developments that occurred before the actual discovery of electricity. This is certainly true of the first application of electricity to communications, namely, the telegraph, whose predecessors include marathon runners, bonfires built along a chain of hills or mountains, semaphore, and the Pony Express. These forms of communication at a distance were obsolesced by the telegraph, telephone, and eventually, radio, television, telex, cable, satellites, and facsimile (fax).

An interesting footnote to the development of the telegraph is that the original purpose of this invention was to provide communications in order to control the railroad system; hence, the telegraph was not only the beginning of telecommunications but it was one of the first electric cybernetic control devices. It was only after it was installed along the railroad lines that its use as a telecommunications medium for the public was realized.

2. **Automation**: As with telecommunications, the roots of automation and the computer can be found in preelectric mechanical devices. The first computers or computational devices were the abacus and the slide rule. Mechanical adding machines, representing the next level of sophistication in computation, were operated by turning a crank which activated a number of gears. The crank action of these machines was eventually electrified, but the actual computation remained mechanical until the development of the electronic computer. The first attempt at building a computer by Charles Babbage in the nineteenth century involved a totally mechanical design. Arthur Porter built a computer in 1933 using a Meccano set which can still be seen in the Science Museum in London, England. The development of electronics and the possibility of creating logic circuits permitted the construction of the first computers, beginning with the ENIAC in 1946. Other cybernetic control devices led to automation of many manufacturing activities and control processes such as aircraft navigation. But the first cybernetic control devices were mechanical devices found in preelectric factories. In essence, modern automation based on computers combines the principles of mechanical feedback with the flow of electric information. With electricity and electronics, the flow of information was greatly accelerated and control of information could be made much more intricate and accurate.

3. **Information storage**: The information revolution is not solely a product of telecommunication and automation; the quantum jump in the storage of information that electric-configured devices permitted was another critical factor. The preelectric roots

of information storage extend to the very beginning of human communication, in which information was stored in the mind and only later in books and libraries, the storehouses of print. They also include the photograph, the motion picture, and the phonograph record, all of which began as purely mechanical devices but were later electrified. "The phonograph, which owes its origin to the electrical telegraph and the telephone, had not manifested its basically electric form and function until the tape recorder released it from its mechanical trappings" (McLuhan 1964, 275). Electrification, however, did not in principle change the mode of the information storage, but only affected the overall efficiency of these devices. The first true electrification of information storage occurred with magnetic tapes which actually stored information in an electrical form rather than in a graphic, written, printed, or photographic manner. First used to store audio signals, magnetic tape was later upgraded to store video signals and to provide memory storage for computers. Other forms of electronic information storage for computers have since been developed using magnetic disks and bubble memory.

4. General systems theory: One side effect of the electronic communications revolution is general systems theory or the operations research approach. The linear mode of thought characteristic of the industrial age collapsed with the introduction of electricity. The new speedup of information forced an integration of sources of data and knowledge which, during the industrial age, could be compartmentalized. The systems approach, in making the best use of computer hardware, led to the development of software to program tasks for the computer. As with the other breakthroughs, the general systems approach and software have preelectric analogs. The way in which preliterate hunting and gathering cultures maintained a balance with nature represents the origins of the general systems approach. The operations research approach was first deployed in military campaigns dating to Alexander the Great. Software was first employed in the form of programming cards for the Jacquard loom and later with the punched rolls of the player piano.

5. **Microelectronics**: The use of miniaturization as a strategy to enhance communication and information storage predates electricity and includes constructing models or plans as well as developing microfilms, microfiches, and other optical microforms. New patterns emerge and qualitative changes occur with each dramatic change in scale. The first computers built with vacuum tubes were large, clumsy, and extremely expensive. The history of computing has seen a steady reduction in their size. The invention of the transistors which replaced vacuum tubes resulted in more compact and practical computers. The development of transistors was key to the use of computers by large companies, government agencies, and educational institutions. It was the microchip, an integrated circuit etched on a tiny silicon chip, that led to microcomputers and the desktop- or personal-computer revolution. The recent emergence of laptop computers and now electronic notebooks is once again creating a dramatic change in the use of computers.

One important effect of miniaturization is portability. In fact, the original motivation to develop the microelectronic chip was portability: reducing the size of computers to facilitate space travel. The hand calculator is one of the spin-offs of the space program. Increased portability is one of the results of each communication revolution. Speech permitted man to take his experiences from one place to another. Tales and legends also provided a means to store experiences and to move them through time. The written record provided an additional vehicle of portability across both space and time. As the materials upon which man wrote became progressively lighter, evolving from stone to clay tablet to animal parchment to papyrus scrolls to paper books, they also became more portable. Print also increased portability by permitting multiple copies of a book to be made cheaply and quickly. Books no longer had to be chained to a table in the monastery or the library, as circulating libraries came into existence. As the efficiency of printing increased and the costs of books decreased, readers could afford their own collections. Moreover, the disposable paperback book heralded a breakthrough in the dissemination of information through print.

Computers created a new level of portability. Information is easily transported from one computer to another through networking. Diskettes containing data can also be used to transport data or software routines from one site to another. With laptops and notebooks it is possible to carry a computer wherever it is needed. Microelectronics or computer chips have also resulted in the design of customized computing devices for a large variety of equipment and machines ranging from the fuel injection systems of car engines to kitchen appliances.

Contemporary information processing and telecommunications devices have evolved through the integration of the above five technological breakthroughs. These technologies are compatible and reinforce one another's operations. Hybridizing them has produced hundreds of information-processing devices for myriads of social, administrative, political, economical, and educational activities and institutions.

NEW INFORMATION PATTERNS EMERGE AT THE SPEED OF LIGHT

Innis explained the space-time patterns of written communication and socioeconomic interactions in terms of the ease and speed with which written material or information could be transported. These notions are also valid for explaining the changes that occur with the electrification of information. McLuhan noted, as did Einstein, that new information patterns emerge at the speed of light. "Today it is the instant speed of electric information that, for the first time, permits easy recognition of the patterns and the formal contours of change and development" (McLuhan 1964, 352). McLuhan points out how the telegraph, "by providing a wide sweep of instant information," revealed weather patterns that had never been observed before. Information transmission at the speed of light is so nearly instantaneous that the barriers of distance are totally eliminated as far as the flow of information is concerned. "Many analysts have been misled by electric media because of the seeming ability of the media to extend man's spatial powers of organization. Electric media, however, abolish the spatial dimension, rather than enlarge it" (McLuhan 1964, 255).

McLuhan's insight about the speedup of information flow with electricity and telecommunications can be extended to the computer. The computer as an information-processing device repeats the same simpleminded calculations that a human thinker can perform but it can perform them automatically, faster, longer, and more accurately. The quantitative change which results from the automation of information processing results in a qualitative change in which new patterns emerge because more data can be encompassed in a single stroke. The accuracy of the computer modeling that the Club of Rome commissioned to show the "Limits to Growth" are disputed by some, but their impact on making humankind more aware of many of the potential ecological disasters we face was invaluable. This is a perfect example of the power of computers to provide new insights.

FRAGMENTATION IN THE AGE OF LITERACY

Before the age of literacy, the thought patterns and social forms of society were coherent, cohesive, and integrated. With writing, a fragmentation of social and psychic patterns began which reached its zenith in the mechanical forms of the assembly line characteristic of the industrial age. This was particularly true of education, where knowledge was divided into specialized disciplines or subjects which were studied separately and taught by individuals who rarely communicated professionally with other scholars outside their own narrowly defined disciplines. Writing encouraged this fragmentation in two ways: through its direct influence, and through the printing press, which functioned as the precursor of mass production. Writing served as a paradigm for separation and fragmentation. It inspired the organization of information into sundry categories and enumerations. "It has been suggested that writing was invented in Sumer to keep tallies and to make lists, and hence, was an outgrowth of mathematics" (Innis 1972, 26). This observation is further supported by the etymology of the Semitic root for writing, *spr*, which also means to count and to tell. One of the consequences of the closely linked arts of writing and enumeration was the creation of bureaucracies which assisted and gave rise to theocratic governments. "The beginning of the principle 'divide and rule' was evident in recognition of the religions of Babylon and Egypt and encouragement of the religion of the Hebrews at Jerusalem" (Innis 1951, 135). "Priestly bureaucracies" gave way to "military bureaucracies" and the principle of "divide and rule" was expanded to the principle of "divide and conquer." This led to centralized world states fragmented into individual provinces. The written word not only created a new geopolitical environment, it also transformed learning. A division was created between the knower or author and that individual's written knowledge. The divorce between the author and his or her writing permitted the objectivity necessary for scientific thought. "The literate man or society develops the tremendous power of acting in any matter with considerable detachment from the feelings or emotional involvement that a non-literate man or society could experience" (McLuhan 1964, 79). At the heart of the scientific method that writing encourages is the specialization and fragmentation of subjects which can eventually flip into tunnel vision.

Phonetic writing with an alphabetic script reinforced the spirit of fragmentation that writing induced. Each word is subdivided into its basic sound values or phonemes, which are then represented by abstract meaningless visual signs. "The phonetic alphabet was a technical means of severing the spoken word from its aspect of sound and gesture" (McLuhan 1964, 193). The printing press reinforced and enhanced the separation and fragmentation that the phonetic alphabet encouraged. Before, print texts had to be read aloud to be deciphered and understood (McLuhan 1962, 103–5). Vocalizing written texts helped preserve elements of the oral tradition. "The lack of assistance to readers, or of aids to facilitate reference, in ancient books is very remarkable. The separation of words is practically unknown, except very rarely when an inverted comma or dot is used to mark a separation where some ambiguity might exist. Punctuation is often wholly absent, and is never full and systematic" (Kenyon 1937, 65). With the printing press, words are uniformly separated, punctuation is used systematically, texts are regularized, and spelling and grammar are standardized. Readers can read much more quickly and decipher the text silently for themselves. With the advent of silent reading from uniformly manufactured texts, the gap between the visual bias of literacy and the oral tradition widened. This further fragmented many aspects of social, economic, political, psychic, and spiritual life.

The modern scientific revolution which began in the Renaissance and culminated in the breakthroughs of Newtonian mechanics led directly to the Industrial Revolution, mass production, and the assembly line, all of which are based on segmentation and specialization of task. "The restructuring of work and association was shaped by the techniques of fragmentation, that is the essence of machine technology" (McLuhan 1964, 8). "The separation of function and the division of stages, spaces and tasks are characteristic of literate and visual society and of the Western world. These divisions tend to dissolve through the action of the instant and organic interrelations of electricity" (McLuhan 1964, 247).

THE GLOBAL VILLAGE

A key element in McLuhan's historical overview of communications is that electric information moving at the velocity of light creates new patterns of communication and social interactions, which he describes as "an instant implosion" that reverses the specialism of the literary age and contracted the globe to a village in which "everybody lives in the utmost proximity created by our electric involvement in one another's lives" (McLuhan 1964, 35).

Electricity bringing information instantaneously from the four corners of the planet invests distant events with a personal dimension; it is as though they are occurring in one's own community. Communities across the globe become entwined in one another's affairs. Electrically based media transform the role of the individual in society; a social grid of highly independent individuals gives way to tribal patterns of intense involvement with one another and a return to elements of the oral tradition. McLuhan pushes Innis's notion of command over space when considering the impact of telecommunication in which information travels at the speed of light. The command over space reduces the entire globe to the dimensions of a village — a "global village."

"Electricity offers a means of getting in touch with every facet of being at once, like the brain itself. Electricity is only incidentally visual and auditory; it is primarily tactile" (McLuhan 1964, 249). Electricity keeps us in touch with one another whether we are in the same place or separated by physical distances which electric signals jump in an instant.

The introduction of electric information changed our view of work from that of a job to that of a role. A role becomes a way of life and gives more of an identity to an individual than a job. The reversal of mechanical forms by electricity has also affected the world of education and academe, although not to the extent that it has affected the commercial world. Today, there is a more dedicated attempt to integrate studies and make sure that students are acquainted with a wider range of issues and disciplines than was the practice forty years ago. There is far more communication between disciplines and some scholars now pursue interdisciplinary studies. Many universities now offer cross-disciplinary programs. The environmental movement has spawned the new interdisciplinary field of ecology. Other foci of study such as future studies, peace studies, native studies, women's studies, and black studies are examples of interdisciplinary fields that have appeared on university campuses to address the needs of students who have specific social and political concerns.

CENTRALIZATION VERSUS DECENTRALIZATION

In the literate age, centralized control was achieved by the technique of divide and conquer. This model of fragmentation, Innis and McLuhan contended, was a by-product of alphabetic literacy and was reinforced by the mechanical forms of industrialization that the printing press inspired. "Political organizations determined to an important extent by the limitations of armed force and characterized by centralized power emphasized the capital and left their impression on cultural activity. They provided a shelter for the development of communication facilities. Communication was subordinated to the demands of centralized power in religion and in political organization; it was characterized by the use of the eye rather than the ear. The scribe occupied a strategic position in centralized bureaucracies" (Innis 1951, 135). The pattern Innis described for literate society before the printing press applied with even greater force to print-dominated cultures, particularly once the Industrial Revolution commenced.

Centralized patterns of organization, which generally characterize political and religious institutions, extended during the industrial age to forms of work, particularly in manufacturing and distribution. This centralization reverses with the introduction of electricity: "Electric information provides instant data access equally to all members of the organization regardless of their hierarchical positions. For the external action of an organization this fact of instant information spells total decentralization" (McLuhan 1988, 67).

Industrial organization requires collecting staff and supplies at centralized locations for manufacturing and distribution. In electrically configured society, activity shifts from manufacturing material goods to processing information, which is a decentralized activity. The mode of organization is like that of the electric circuit: no point along a circuit (or a circle) may be regarded as the unique center of the circuit; each point is as central and essential as the next. The telephone, for example, places us at the center of a worldwide network of information. Computers linked by telephone lines permit access to vast stores of knowledge from any point on earth. The microelectronic revolution, which has greatly reduced the cost of electronic equipment and increased portability through miniaturization, enhances this trend towards decentralization. There is little point in going to a centralized office to use information equipment that one can operate just as easily at home. One can network with others by telephone as easily from home as from the office.

Innis did not share McLuhan's notion that electricity decentralizes communication and other socioeconomic patterns. This divergence of views represents perhaps the most serious dichotomy between the two scholars. Innis saw radio as a force that promotes centralization. "The radio appealed to vast areas, overcame the division between classes in its escape from literacy, and favoured centralized bureaucracy" (Innis 1951, 82). McLuhan disagreed with this assessment. "A good example of this technological blindness (i.e., blindness to the effects of one's most significant forms of invention) in Innis himself was his mistake in regarding radio and electric technology as a further extension of the patterns of mechanical technology" (Innis 1972, xii, McLuhan's foreword).

Is McLuhan's critique of Innis fair? In terms of the centralized broadcast mode of operation of radio and television during the time that Innis and McLuhan were active, it is not. The centralized mode of operation of radio and television was dictated by the economics of a limited number of channels. The emergence of cable TV, fiber optics, communication satellites, pay TV, videotext, and eventually, the information superhighway will vastly increase the number of channels and services available. This will lead to the decentralization of radio and television operations. Television regulatory agencies are already experiencing difficulties in regulating this once highly centralized industry because of the proliferation of new communication channels. McLuhan's critique is therefore valid in the long term because the present centralized broadcast mode of radio and television will not dominate these media forever. The way that cable TV has cut into the monopoly of the networks is a perfect illustration of this point.

The conflict between Innis and McLuhan over the issue of centralization is due to a difference in emphasis. Innis focused on the hardware aspects of electronic communication, which is, or has been, centralized, whereas McLuhan was more concerned with the software or information flow, which is decentralized. Consider the telephone system as an example. The hardware consists of the telephones, the wires connecting them, and the central switching stations. The software or information the system carries is provided by those who use the system to communicate with each other. The hardware may be centralized, as was dramatically demonstrated by General Wojciech Jaruzelsky when he pulled the plug on the telephone system during the suppression of the Solidarity movement in Poland in December 1981. The information flow on the telephone system is, on the other hand, decentralized. A phone call can be initiated from any point in the system. In the long run, not even centralized control of the communications hardware, the army, and the police can support a totalitarian regime, as the dramatic collapse of communism and other forms of dictatorship have demonstrated.

A similar centralization of hardware and a decentralization of information is true of a computer network, in which there is a central computer and an information bank. A network of personalized microcomputers, or the Internet, the network of networks, on the other hand, is totally decentralized. The revolutionary effect of microcomputers in education will be a result of their decentralized grass-roots use. Centralized mainframe computers in use at least ten years before the introduction of microcomputers had very little effect on schools.

HARDWARE VERSUS SOFTWARE

There has been a dramatic shift in economic activity from an almost total emphasis on hardware at the height of the industrial age to a focus on information and software products. The movement of information is now the main source of wealth in the computer age. These observations are easily confirmed: look at employment levels in the different sectors of our economy. Before the Industrial Revolution, most workers were employed in agriculture. After the Industrial Revolution, the number in manufacturing soared, those in the service sector remained steady, and the number in agriculture declined dramatically to the present level of 5 percent of the North American work force. Today, the number of information handlers has grown dramatically to approximately 50 percent of the work force. Still more growth is expected. Those in manufacturing, currently 20 percent of the work force, will decline further as the full impact of robotization and automation transforms industrial activity.

HYBRID SYSTEMS

"The hybrid or the meeting of two media is a moment of truth and revelation from which new form is born" (McLuhan 1964, 55). Hybrid systems operate more efficiently than the individual components from which they emerge. Hence, the synergy they represent releases new forms of energy, creating additional "vortices of power." Electric systems, which favor integration, give rise naturally to hybrid systems.

The hybridization of the computer, the video screen (or CRT), and microelectronics have produced the low-cost personal computer and now the highly portable electronic notebook. The modem, which permits the marriage of the computer and the telephone system, creates instantaneously vast computer networks, such as the Internet, linking the databases of the world to the home or office. The marriage of the TV broadcast system with interactive intelligent home terminals and databases creates videotext systems such as Prestel in England, Antiope in France, and Telidon in Canada.

SOCIETIES IMITATE THEIR TECHNOLOGIES

McLuhan observed that the fragmentation and centralization of industrial or mechanical society was reversed in the electric age of information to systems of integration and decentralization which retrieved some aspects of the social organization of preliterate society. These observations were based on the following set of assumptions:

1. The dominant tools or technologies of a society create patterns of usage which infiltrate or penetrate the social structures of a society.
2. These patterns change those structures.
3. Eventually, the social structures come to imitate or replay the patterns by which these dominant technologies are organized.

Communities alter their form to accommodate the technologies they use, which literally become extensions of themselves. The social structure so amended by a technology tends to reproduce new technological innovations in the same mold, thus reinforcing the existing pattern. "Once a new technology comes into a social milieu it cannot cease to penetrate that milieu until every institution is saturated" (McLuhan 1964, 177).

Eventually, the school system will be penetrated by the milieu of the computer age, one of the last institutions to make the transition. Educators and academics pride themselves on their ability to incorporate the latest in new ideas, which is true as far as their discussions and scholarly work are concerned. Unfortunately, this does not always translate into the way they organize the educational activities of their students.

There is an analogy between McLuhan's observation that societies imitate their technologies and Thomas Kuhn's notion that once a new scientific theory scores a success, it serves as a paradigm for further scientific work, and every conceivable extension, articulation, and application of the original idea is developed (Kuhn 1972, 23–42). This process, which he calls "normal science," occurs with technology, as well. A technological breakthrough also serves as a paradigm which is extended, articulated, and applied in every conceivable way. For example, the idea of building things from repeatable fragmented identical elements began with the introduction of the phonetic alphabet. As we will see, the extension of this idea resulted in the development of codified law, monotheism, abstract science, and deductive logic. The articulation of the alphabetic paradigm of the repeatability of fragmented identical items in the hardware realm of mechanics created the printing press, the assembly line, mass production, and the general organizational principle of the industrial age.

The evolution of this fragmented, specialist, linear, sequential mode of organization came to an end with the introduction of electricity and its application to information systems. Slowly the patterns of industrial organization and social interaction reversed themselves as electricity reconfigured social structures. This development continues with the penetration of microelectronics or integrated circuits into organizational structures, which in turn will lead to a continuation of the integration and decentralization of social structures along with a greater shift to software and hybrid systems.

In *The Third Wave*, Alvin Toffler discusses the possibility of being able to work at home by making use of an electronic workstation and increased communication with the outside world. This notion of an "electronic cottage" incorporates three major trends of economic life in the age of electric information:

1. the decentralization of economic activity,
2. the integration of work with home life and leisure activities, and
3. the replacement of a job with a role in society.

BREAKPOINT BOUNDARIES

The boundaries between two technological orders and their subsequent social patterns have been described by Kenneth Boulding as "'break boundaries' at which [a] system suddenly changes into another or passes some point of no return in its dynamic processes" (Boulding 1961).

Identifying break boundaries is one of the methods used by the Toronto School of Communication to understand the dynamics of social systems and their interactions with media of communication and information-processing technologies. The major break boundaries in communications and informatics are those associated with the introduction of ideographic writing, mathematics, phonetic writing, the phonetic alphabet, abstract science, the printing press, telecommunication, automation, computers, magnetic storage of information, general systems theory, and microelectronics. Each of these innovations created its own breaks in social patterns.

ACOUSTIC VERSUS VISUAL SPACE

The separation between the oral and literate traditions represents a break boundary for Innis and McLuhan. These two traditions are characterized by their respective oral and visual biases of communication and perception. McLuhan studied how the dominant mode of communication in a culture created a sensory bias or grid by which the rest of the world was viewed. He created the notion of acoustic space and visual space, two worlds in which oral and literate society, respectively, lived. "Rural Africans live largely in a world of sound — a world loaded with direct personal significance for the hearer — whereas the Western European lives much more in a visual world which is on the whole indifferent to him" (McLuhan 1962, 28). McLuhan's observation is echoed by Innis: "The oral tradition implies the spirit but writing and printing are inherently materialistic" (Innis 1951, 130). "The oral discussion inherently involves personal contact and a consideration for the feelings of others, and it is in sharp contrast with the cruelty of mechanized communication" (Innis 1951, 199).

Acoustic space not only includes the aural and the oral but also the tactile, and the interplay of these senses. It is a space characterized by direct contact with reality. Thought patterns within this world are concrete and much use of metaphor and analogy is made. The visual world is one of abstraction and deductive logic. Geometry and science characterize visual space, whereas art and music are more representative of acoustic space. As noted earlier, the perceptual bias of visual space is linear, sequential, rational, fragmented, causal, abstract, and specialized, whereas that of acoustic space is simultaneous, concrete, intuitive, all-embracing, mystical, inductive, and experiential. Visual space is characterized with patterns associated with the specialization of the left hemisphere of the brain, acoustic space with those of the right hemisphere.

The visual and oral bias of writing and speech, respectively, are obvious. The bias of electric communication is more difficult to identify. Electric information patterns, however, reflect a number of the characteristics of oral communication, namely, integration, simultaneity, despecialization, decentralization, and less emphasis on hardware. The parallels between the oral tradition and electric information patterns led McLuhan to the conclusion that the inherent bias of electric communication is oral (McLuhan 1962, 86). "A speed-up, such as occurs with electricity, may serve to restore a tribal pattern of intense involvement such as took place with the introduction of radio in Europe, and is now tending to happen as a result of TV in America. Specialist technologies detribalize. The non-specialist electric technology retribalizes" (McLuhan 1964, 24).

MULTIDISCIPLINARITY VERSUS SPECIALIZATION

McLuhan recognized and valued a generalist and multidisciplinary approach. The depth and learning of the expert is valued as long as it does not exclude the learning and ideas of other experts. The tunnel vision of the specialist, however, is avoided. "The specialist is one who never makes small mistakes, while moving toward a grand fallacy" (McLuhan 1964, 124). The multidisciplinary approach is the only way to get at the large patterns of the interplay between a society and its technologies. It is an approach that blends naturally into the Innis–McLuhan style of scholarship. "Any subject taken to depth at once relates to other subjects" (McLuhan 1964, 347).

The effects of electric information patterns naturally reinforce the multidisciplinary approach of Innis and McLuhan. Electric technology "ends the old dichotomies between culture and technology, between art and commerce and between work and leisure" (McLuhan 1964, 35, 346).

Innis also found these dichotomies artificial and a barrier to meaningful scholarship. He described one of his essays as "a critical review, from the points of view of an historian, a philosopher and a sociologist, of the structural and moral changes produced in modern society by scientific and technological advance" (Innis 1951, 190). Innis and McLuhan adopted multidisciplinary approaches in order to deal with questions that were most relevant and meaningful to humankind. They saw overspecialization as a major stumbling block preventing scholars from dealing with pressing philosophical questions.

THE MYTH OF OBJECTIVITY

Innis and McLuhan challenged many of the biases of visually oriented, left-brained specialists, including their belief in the existence of a single objective reality. "An interest in the bias of other civilizations may in itself suggest a bias of our own" (Innis 1951, 33). According to McLuhan, new media affect the sense ratios or perceptions of their users, and hence shift their perceptual realities, creating multitudes of realities. The objective reality of Einstein, who lived in the electric age, was quite different from that of Newton, as is reflected in the differences in their physics. Thomas Kuhn in *The Structure of Scientific Revolutions* articulates the process whereby objective realities shift each time the current phase of "normal science" is replaced by "revolutionary science." Jacques Lusseyran, who was blinded as a youth, criticizes the "myth of objectivity" in moving terms based on his own experience:

> When I came upon the myth of objectivity in certain modern thinkers, it made me angry. So there was only one world for these people, the same for everyone. From my own experience I knew very well that it was enough to take from a man a memory here, an association there, to deprive him of hearing or sight, for the world to undergo immediate transformation, and for another world, entirely different but entirely coherent, to be born. Another world? Not really. The same world rather, but seen from another angle, and counted in entirely new measures. When this happened, all the hierarchies they called objective were turned upside down, scattered to the four winds, not even theories but like whims. (Lusseyran 1963, 112)

THE USER IS THE CONTENT

Operating under the "myth of objectivity," many scholars believe that they can analyze a text or work of art solely on the basis of its content, without any reference to the culture in which it is set, the intentions of the person who created it, or the attitudes of the audience for which it was intended. They also labor under the misconception that they can carefully disentangle their own cultural biases and read and analyze a text in a completely objective manner. The ambiguities of language or the author's intention are often ignored by these scholars. The text, play, movie, or art object is analyzed as a figure without a ground. What these scholars fail to understand is that they can never escape the bias of their language, their culture, and their particular form of scholarship.

Innis warned us of this problem in *The Bias of Communication*: "We must all be aware of the extraordinary, perhaps insuperable, difficulty of assessing the quality of a culture of which we are a part or assessing the quality of a culture of which we are not a part. In using other cultures as mirrors in which we may see our own culture we are affected by the astigma of our own eyesight and the defect of the mirror, with the result that we are apt to see nothing in other cultures but the virtues of our own" (Innis 1951, 132). Each reader or viewer brings his or her own experience to a medium and transforms the content according to his or her own need, which McLuhan expressed with his famous one-liner, "the user is the content."

Teachers and managers who were trained before the advent of computers cannot fully comprehend the revolution through which we are passing because they view it through the eyes of a past era. The true integration of computing into education and the workplace will only occur when people who are now growing up with the technology become the teachers and managers of the future. Nevertheless, we, the members of the transitional generation who learned how to use computers as adults, can make the transition from the industrial age to the computer age easier by realizing we must come to grips with a new phenomenon for which we were never trained.

ART AS RADAR AND AN EARLY WARNING SYSTEM

One of the unique aspects of the Innis–McLuhan approach is the way that the insights and methods of artists are integrated into media studies. Not only is this a natural extension of the multidisciplinary approach, it is also a reflection of their belief in the superiority of artistic sensibility and its ability to detect currents of change before they fully impact on a society. McLuhan was fond of quoting Wyndham Lewis, who wrote: "The artist is always engaged in writing a detailed history of the future because he is the only person aware of the nature of the present." McLuhan believed that the artist's insights were like radar or an early warning system which could pick up the social and cultural problems and challenges that new technologies created.

Just as Innis (Innis 1972, xi) had used "historical situations as a lab in which to test the character of technology in the shaping of cultures," so McLuhan used artistic insights — his own and those of others — as a lab in which to study the effects of technology. As a literary critic, he naturally was more concerned with the arts than was Innis, the economic historian. But Innis was no less an artist. "The artist is the man in any field, scientific or humanistic, who grasps the implications of his actions and of new knowledge in his own time. He is a man of integral awareness" (McLuhan 1964, 65).

There are critics who have tried to dismiss the work of Innis and McLuhan as that of poets. As we see from the above definition of an artist, this is far from a negative criticism; rather, it is a vote of confidence. In terms of the language of the split brain, the integration of scholarship and art represents an attempt to integrate the left and right hemisphere patterns of knowing. Art education will be one of the important tools in helping our school system make the transition from the industrial age to the information age.

THE ORAL TRADITION AND PROBES

One of the paradoxes of the Innis–McLuhan approach is that while both men were extremely well-read scholars who wrote extensively, they were in many ways more a part of the oral tradition than the literary one. They were both excellent teachers whose lectures were a great source of delight and insight for their students. They were both masters of the spoken word more than of the written word. Both were criticized for their difficult prose. Innis admits his bias in an essay entitled "A Critical Review": "My bias is with the oral tradition particularly as reflected in Greek civilization, and with the necessity of recapturing something of its spirit. For that purpose we should try to understand something of the importance of life or of the living tradition, which is peculiar to the oral as against the mechanized tradition" (Innis 1951, 190). McLuhan also confessed to a similar bias: "I have to engage in endless dialogue before I write. I want to talk a subject over and over" (Schickel 1965). He believed that in the age of electric information, the one-liner was the optimal mode of communication. McLuhan's memorable one-liners — the medium is the message, the global village, the Gutenberg Galaxy, the Mechanical Bride, and the user is the content — have enriched the English language.

Further evidence of McLuhan's bias towards the oral is his use of what he called "probes" for exploring new ideas in a way reminiscent of Plato's dialogues. McLuhan had no compunction about tossing out an untested or unfinished idea and playing with it to see what could be learned. He often remarked that "even if a probe is half true that is a lot of truth." His method was a very useful tool, but it confounded the academic establishment, which found it difficult to condone inaccuracy even if it led to enlightened insights. McLuhan did not restrict his use of probes to discussions, but published them in his many books, a sin many academics could never forgive.

AN ANTIACADEMIC BIAS

The unorthodox methodology of Innis and McLuhan had an antiacademic bias built into it. The two theorists were highly critical of the academic approach to understanding the relationship between society and its media and technologies. Innis wrote: "Finally we must keep in mind the limited role of universities and recall the comment that 'the whole history of science is a history of the resistance of academics and universities to the progress of knowledge'" (Innis 1951, 194).

One of the specific critiques Innis and McLuhan made of academics is that they are too specialized, and hence unable to deal with the complexity and multidisciplinary aspects of communications. "In education the conventional division of the curriculum into subjects is already as outdated as the medieval trivium and quadrivium after the Renaissance" (McLuhan 1964, 124). Another of their concerns was that the human dimension of academic discussions or scholarship was often missing. "Perhaps we might end by a plea for consideration of the role of the oral tradition as a basis for a revival of effective vital discussion and in this for an appreciation on the part of universities of the fact that teachers and students are still living and human" (Innis 1951, 32).

MONOPOLIES OF KNOWLEDGE

One of the reasons that Innis was critical of universities is that they, like other literary-based institutions, represent what he called a "monopoly of knowledge." He writes: "A complex system of writing becomes the possession of a special class and tends to support aristocracies" (Innis 1951, 4). The academic style of writing, which avoids using vernacular expressions in an attempt to sound erudite, merely narrows the audience that can share its knowledge.

The first monopolies of knowledge were formed by the priestly bureaucracies of Babylon and Egypt. Secret learning societies in the ancient Greek and Roman world also maintained such monopolies. During the Middle Ages, "monopolies of knowledge controlled by monasteries were followed by monopolies of knowledge controlled by copyist guilds in the large cities" (Innis 1951, 53).

The printing press broke up these monopolies and enabled a new one to evolve. "Freedom of the press has been regarded as a great bulwark of our civilization and it would be dangerous to say that it has become the great bulwark of monopolies of the press" (Innis 1951, 139). "Even science with its emphasis on a common vernacular and on translations has come under the influence of monopolies of knowledge in patents, secret processes and military security measures" (Innis 1951, 129). It is probably the maintenance of a monopoly that makes the defenders of an old school of scientific thought so tenaciously defend their theories from the attack of the proponents of a new paradigm. Physicians are another group who have maintained a monopoly of knowledge through often mystifying patients with unnecessarily complex terminology and the use of Latin to write prescriptions.

As we have seen, there is a general trend with electricity to speed up the flow of information and decentralize its sources, which break up these monopolies and make secrecy difficult. "Mechanization has emphasized complexity and confusion; it has been responsible for monopolies in the field of knowledge; and it becomes extremely important to any civilization, if it is not to succumb to the influence of this monopoly of knowledge, to make some critical survey and report. The conditions of the freedom of thought are in danger of being destroyed by science, technology, and the mechanization of knowledge, and with them, Western civilization" (Innis 1951, 32).

The fears that Innis expressed in 1951 have disappeared because automation and connectivity have led to open systems in which information is free to flow and small elites are no longer as able to control the creation of wealth. This new environment, however, leads to new challenges, which we will explore in later chapters. For the moment, we turn to an examination of the evolution of language and the information systems that underlie our economic and political institutions and our education system.

3

FROM THE SPOKEN WORD TO COMPUTING: The Evolution of Language

THE FIVE MODES OF LANGUAGE

The computer is the most recent technique for organizing human thought in a long series of techniques and technologies, beginning with speech, for communicating, storing, retrieving, organizing, and processing information. The series includes spoken language, pictures, tallies, clay tokens, picture writing, logographic (pictographic or ideographic) writing, syllabaries, the alphabet, abstract numbers, numerals, mathematical signs (+,-,=), the concept of zero, geometry, mathematics, logic, abstract science, maps, graphs, charts, libraries, the printing press, encyclopedia, dictionaries, bookkeeping techniques, the scientific method, photography, the telegraph, the telephone, cinema, radio, audio recording, television, video recording, optical disks, computers, control theory, and cybernetics.

Computing, however, is more than a new technology. It represents a new form of language, if we accept that language is defined as a system for both communications and informatics. Computing is part of an evolutionary chain of languages which also includes speech, writing, mathematics, and science. In order to establish the hypothesis, I will show that these five modes of language are distinct, each with its own semantics and syntax. Through historical analysis, I will demonstrate that each new mode of language evolved from the previous forms as new information-processing needs arose and that each new language subsumed the structures and elements of the earlier languages.

Exploring the origins and evolution of language helps us understand how it has affected the development of the three domains of cognition, communication, and education. My definition of language as a medium for both communication and cognition corresponds to Lev Vygotsky's idea that language is a vehicle for "social intercourse" and "generalizing thought" (Vygotsky 1962). Formal education may therefore be viewed as the learning of different languages. Before starting school, one begins to learn the oral form of one's mother tongue. In school, one learns the written forms of the vernacular tongue and the language of mathematics, known colloquially as the three Rs. Students then learn the language of science and possibly a foreign language. "The influence of scientific concepts on the development of the child is analogous to the effect of learning a foreign language, a process which is conscious and deliberate from the start" (Vygotsky 1962, 109). If this is the case, then the optimal time for a child to learn the languages of mathematics, science, and computing is the age when it is easiest for them to learn a new foreign language: five to six years.

With the advent of microcomputers, education now includes the study of a new distinct language, the fifth language — computing. Essentially, education becomes the study of the five verbal languages of speech, writing, mathematics, science, and computing. A comprehensive education must also include nonverbal languages and activities such as the visual arts, music, and physical education, but for this study I will focus on the integration of the five verbal languages and their impact on education and work.

A New Concept of Language

The use of computers to communicate and process information represents a new language with which educators must deal as they have with the languages of mathematics and science. To understand the impact computers have on the way we organize information and work and educate our children, it is necessary to recognize not only that computing is a new form of language but also that it changes our very notion of what defines or constitutes a language. Language is communication plus informatics. This definition may be unorthodox from a linguist's point of view, but it will prove extremely useful for our discussion of education.

Traditionally, the sole focus of linguists in defining language has been on communication. One example is Edward Sapir's definition of language: "A purely human and non-instinctive method of communicating ideas, emotions, and desires by means of a system of voluntarily produced symbols" (Sapir 1921). While it is true that the main function of language has been communication, this is not its only function; language also plays a key role in the processing of information, including its storage, retrieval, and organization. Language is a tool for developing new concepts and ideas; it is an open-ended system. Since writing, mathematics, science, and computing permit the development of ideas that could never arise through the use of speech alone, we must consider these other modes as distinct, albeit related, languages.

While the information-processing aspects of language became more apparent with the introduction of writing, spoken language also served as a system to store, organize, and process information. Ernst Fisher suggests that human communication evolved differently from that of animals because man's activities required that information be processed. "The development towards work [i.e., using tools to change one's environment] demands a system of new means of expression and communication that would go far beyond the few primitive signs known to the animal world. Animals have little to communicate with each other. Their language is instinctive: a rudimentary system of signs for danger, mating, etc. Only in work and through work do living beings have much to say to one another" (Fisher 1963, 23). At times, animals may communicate with other members of their species the way humans do, but they cannot use language to process information or to create new information.

Generalizing and extending Sapir's definition, I define language as: "A purely human and noninstinctive method of communicating ideas, emotions, and desires *as well as processing, storing, retrieving, and organizing information* by means of a system of voluntarily produced symbols." Given this definition of language, it is necessary to recognize that speech is not the only form of language. My suggestion that speech and writing, for instance, are two distinct but related forms or modes of language differs from the beliefs of traditional linguists who consider speech as the primary form of language and writing as merely a system for transcribing or recording it. This tradition dates back to Aristotle: "The sounds . . . are symbols of ideas evoked in the soul and writing is a symbol of the sounds" (Bandle et al. 1958, 95).

Ferdinand de Saussure, one of the founders of the field of linguistics, formulated the modern attitude of linguistics towards writing when he wrote: "Language and writing are two different systems of signs; the latter only exists for the purpose of representing the former . . . The subject matter of linguistics is not the connection between the written and spoken word, but only the latter, the spoken word is its subject" (de Saussure 1967). De Saussure's position was reinforced by another immensely influential linguist, Leonard Bloomfield, who wrote: "Writing is not language, but merely a way of recording language by means of visible marks" (Bloomfield 1933, 21).

The relationship between writing and language posited by de Saussure and Bloomfield is far too confining for understanding the implications of microcomputers for the future of communication, work, and education. Their definitions are restricted to a model in which the sole purpose of language is communication; they do not take into account the information-processing capabilities of language, a key consideration for understanding the role of computers in society. These traditional linguists are mired in the industrial age attitude of assigning only a single task to any activity. The definitions of Bloomfield and de Saussure are being reevaluated by a modern generation of linguists who view language from the perspective of the information age, and hence understand the multi-tasking nature of language.

Michael Stubbs critiques Bloomfield's definition of writing as follows: "Writing is not *merely* a record . . . I know from personal experience that formulating ideas in written language changes those ideas and produces new ones" (Stubbs 1982). Joyce Hertzler concurs: "People often find that their thoughts are clarified and systematized, and that necessary qualifications and extensions appear, when they subject them to the more rigorous test of exactness and completeness demanded by the written form" (Hertzler 1965, 444). Frank Smith agrees: "Ideas develop from interaction and dialogue . . . especially with one's own writing" (F. Smith 1982, 204).

While it is true that written language is derived from spoken language, it is useful to regard them as two separate language modes because they process information so differently. Arguments that support the notion that writing is a separate mode of language can also be made for mathematics, science, and computing. These three additional modes of language each have unique strategies for communicating, storing, retrieving, organizing, and processing of information. I have therefore extended the notion of those linguists who consider speech and writing as separate modes of language to claim that speech, writing, mathematics, science, and computing are five separate modes of language which are distinct but interdependent. They form an evolutionary sequence in which the later modes are derived from and incorporate elements of the earlier modes of language. They form a nested set of languages in which the later forms contain all of the elements of the earlier forms.

Speech, the first form of human language, is the basis of all other linguistic modes of communication and information processing. We can define spoken language as the sum of information uttered by human speakers. Only the nonverbal forms of human expression, such as music, dance, and the visual arts, do not derive directly from speech. A picture or a piece of music can describe a sunset nonlinguistically, but a linguistic description is more amenable to information processing because of the abstractness of the description.

Written language, which is derived from speech, is defined as the sum of information which has been notated with visual signs. It differs from speech in that it involves a permanent record, whereas speech disappears immediately after it is uttered. We shall distinguish four different modes or forms of written language: writing (or literature), mathematics, abstract science, and computing. Writing and mathematical notations were the first forms of written language; both grew out of the system of recording payments of agricultural tributes using clay accounting tokens in Sumer just over 5,000 years ago. The language of science and its methodology emerged from writing and mathematics in ancient Greece some 2,500 years ago. The methods and findings of science are expressed in the languages of writing and mathematics, but science may be regarded as a separate form of language because it has a unique way of systematically processing, storing, retrieving, and organizing information which is quite different from either literature or mathematics. Finally, less than fifty years ago, the latest system for processing information emerged from science and mathematics in the form of computing, with its own unique cybernetically based and automated methods for processing and organizing information.

Whether these five modes of information processing and communication should be regarded as separate languages or whether they are merely five different aspects of the human capacity for languaging are questions I will not attempt to answer. For the purposes of this analysis, I will consider speech, writing, mathematics, science, and computing as five distinct modes of language which form an evolutionary chain of development. What these modes of language share is a distinct method for the communication and the processing of information each of which changes our worldview. Just as Sapir argued that cultures dissect nature along the lines dictated by their native languages, so the five languages of information processing each provide a unique framework for viewing the world.

A Model for the Evolution of Thought from Language

In *Thought and Language*, Lev Vygotsky shows that language plays both a communication and an information-processing role. In his attempt to demonstrate the relationship between thought and language, Vygotsky posits three phases in a child's development of verbal thought: "social speech," "egocentric speech," and "inner speech."

The first form of speech — "social speech" — is purely for the purpose of communication and is nonintellectual. After using language as a tool for communication and social intercourse, the child discovers that language is also useful for facilitating his or her thought processes. The child then begins to display a phenomenon known as "egocentric speech" in which the youngster is basically talking to him or herself. The child vocalizes but is not addressing anyone in particular. This form of speech, which commences at about age three, continues to about age seven, when it suddenly disappears and is replaced by "inner speech" — or thought. At first, egocentric speech does not differ greatly from social speech, but as the child matures, the speech takes on more and more of the aspects of inner speech. Vygotsky's observations of children ages three to seven reveal that they resort to egocentric speech whenever they are confronted with a puzzling situation they need to think through. Vygotsky concludes that egocentric speech is basically the child thinking out loud, and that it naturally evolves into inner speech once the child realizes that the vocalization is not necessary for the main function of this form of speech, namely, problem solving or thinking. "We came to the conclusion that inner speech develops through a slow accumulation of functional and structural changes, that it branches off from the child's external speech simultaneously with the differentiation of the social and the egocentric functions of speech, and finally that the speech structures mastered by the child become the basic structures of his thinking" (Vygotsky 1962, 51).

Vygotsky's model is based on the notion that speech or language has two components: communications and informatics. Language in the form of social speech in children younger than age three is pure communication and social interaction. With egocentric and inner speech, language becomes a tool for processing information or assisting the child to think. The only difference between egocentric and inner speech is that the former, which appears first, is vocalized, and the latter is not. But both serve the same function. Once inner speech emerges and is used solely for the purposes of thought, egocentric speech disappears and vocalized speech is used exclusively for communication purposes except in times of stress, when even adults sometimes think out loud by talking to themselves.

Assuming that "ontogeny recapitulates philogeny," that is, that the development of the individual organism of a species follows the same path as the evolution of that species, we may conclude that a similar process occurred in the development of human speech and thought and that there existed a period in human society of purely nonverbal thought and nonintellectual speech. There then followed the integration of thought and verbal communication similar to that described for the child by Vygotsky. Speech took on both a communications and informatics role, although the communications role dominated at first. Admittedly, this conclusion is speculative, yet it does suggest a plausible model for the way in which the informatics aspects of language evolved from the originally purely communicative role of early human language.

The Evolution of the Five Modes of Language

The development of language is the one characteristic of Homo sapiens that distinguishes it from all other species. All species communicate, as the behavior of bees, birds, and primates demonstrates. No other species, however, has created such a perfect tool for abstract thinking as language. The evolution of speech seems to be connected with the use and manufacture of tools and the growth of the brain-case. "Biological evolution ties in with the growth of culture, since the use of language seems obviously associated with habitual tool use (limiting the use of gestures) and with increasing human intelligence" (Vygotsky 1962). The primary use of speech in preliterate societies is for social interactions and the coordination of activities that require cooperation, such as hunting or food gathering.

Spoken language evolved more complex functions and was used in the cultural apparatus of a society to tell tales and sing songs. Eventually, speech was used as a medium to record (or store) and retrieve cultural information in the form of poems, folktales, and folk songs. As stories became more complex, speech was used to organize the information stored in these formats. Organizational forms such as rhyme, rhythm, meter, and plot, in turn, became information tools which permitted larger amounts of data to be stored and successfully retrieved (Havelock 1963).

Eventually, however, speech and the human capacity for memorization encountered limits as to how much data could be recorded in this manner. Writing systems and numerical notations emerged which allowed the amount and type of data being stored to expand enormously. The invention of writing and mathematical notations also had a tremendous impact on the informatic capacity of human language and thought. Written records gave rise to new forms of classification, analysis, and other forms of information processing. The increase in the information-processing capacity that mathematical notations make possible is easily confirmed by comparing the complexity of mathematical calculations that can be done with pencil and paper compared with calculations done solely in one's head.

The increase in the amount and sophistication of data that writing and mathematics made possible eventually gave rise to a new form of language and information processing — the language of science. Scientific activity, whether it is concrete or abstract, is confined to literate societies. Science is not just the gathering of new knowledge about nature; it also consists of giving a shape to that knowledge by organizing it systematically. The effective storage and information processing that writing and mathematics made possible allowed scholars to gather and collect so much data that the only way to deal with the complexity and the information overload that ensued was to develop a new mode of organization known as the scientific method.

Science and the scientific method, however, also became a tool for generating still more information-gathering activities. The information overload that modern science and the scientific method helped to generate became so great in the twentieth century that it led to the development of computers as a way to help scientists cope with the enormous amounts of data they accumulated and the complex calculations they needed to execute.

Vygotsky's work shows how children discover that language, a medium of communication, can also be used as a tool to process information and solve problems. One can easily extend this notion backwards in time and assume that at some point, humans discovered that their system of vocalized verbal signals could be used internally, as a tool for thinking. The extension of speech and its concretization in the form of writing, mathematics, science, and computing levered language as a tool for thinking and amplified its informatics capacity while preserving its communications function. The motivation for the emergence of new forms of language, however, seems to have been strictly the need for a greater informatics capacity, not increased communications.

Writing was first used not for communication but for the keeping of accounting records. For the period just after the emergence of writing, "few literary documents [were] excavated, although the same period has yielded tens of thousands of economic and administrative tablets and hundreds of votive inscriptions" (Kramer 1959). The very first words to be assigned a written form in the Sumerian language were the words for agricultural commodities that were collected as tribute by the priesthood who ran the irrigation system. They used writing to record who paid their taxes to the state. Written numerals were invented at the same time as writing to keep track of the amounts of each commodity that were given in tribute.

The invention of writing and abstract numerals illustrates cyberneticist Ross Ashby's "theorem of requisite variety" (Ashby 1957) as well as the notion that necessity is the mother of invention. According to Ashby, managers can only control a system if they can create a model of it which contains the requisite variety or complexity to accurately describe it. The priests running the irrigation system needed to collect tribute from farmers in order to feed the irrigation workers. They therefore needed to store and keep track of a complex set of data. Given that human memory has difficulty coping with more than seven, plus or minus two, elements at a time, the only way the priests could remember who had paid their tribute and who had not was to create permanent records of the tributes. It was only after the invention of writing for the purposes of economic control that writing was also used for the purposes of communication, and eventually, for other informatic applications such as the composition of stories or written poems.

The development of science was also motivated by purely informatic considerations. Abstract science permitted greater control of nature through better organization of information and the ability to make predictions. The scientific method provided scientists with the requisite variety to control a body of knowledge that the languages of writing and mathematics had not been able to manage.

The invention of computing is still another example of informatics driving the development of language. Without computing, natural and social scientists would not have been able to manage the information overload created by their disciplines. It was only after its initial application as an informatics device that computing was also used for communications, and hence, its name in English is "computer" (as in calculator) and not "word processor," even though far more users process words with computers than compute or calculate with them.

Starting with the ability to record ideas through writing and mathematical notation, human thought has become increasingly more complex. The need to model more complex phenomena has driven the development of the five modes of language. Consequently, each new mode of language is informatically more powerful than its predecessors, but at the same time a little less poetic. Our model of the evolution of language is one in which the information-processing capacity of language becomes more and more important as the complexity of human thought increases. It is essential to remember, however, that all forms of language possess a dual capacity for communication and information processing. Computers are also communication devices and the spoken word has an informatics capacity. Both of these features of language must be addressed when we consider the nature of work and education.

The Semantics and Syntax of the Five Modes of Language

The claim made here that writing, mathematics, science, and computing are distinctive modes of human language is based on the notion that a language is defined by both its information-processing powers and its capacity for communication. To strengthen the claim that these four modes of language may be regarded as languages in their own right and not just derivatives of speech, I will demonstrate that they qualify for this distinction solely on the basis of the criteria established by traditional linguists. According to two of them, Paivio and Begg, "semantics and syntax — meaning and grammatical patterning — are the indispensable core attributes of any human language" (Paivio and Begg 1981, 25). Semantics is the relationship between linguistic signals and their meaning or, in other words, a lexicon. "Naming is undoubtedly the most straightforward and dramatic example of such semantic behavior" (Paivio and Begg 1981, 25). Syntax is the structure or relationship among linguistic signals.

New modes of language evolved to model increasingly more complex phenomena, and hence, according to Ashby's theorem of requisite variety (Ashby 1957), they required more complex structures to function. We therefore expect the semantics and the syntax of the new forms of language to retain the older structures and add their own new unique elements to those structures.

If writing, mathematics, science, and computing are distinct modes of human language which deserve to be differentiated from speech, then they must have distinct semantical and syntactical features above and beyond those of speech. In the case of writing, the semantics of the written word are quite similar to those of the spoken word, although there are examples where a construction which is acceptable in oral language is not valid in prose and vice versa. Oral contractions such as "don't" or "can't" are not widely used in formal prose writing. As a rule, one's written vocabulary is considerably greater than one's oral vocabulary. We often use words in our written communication that we would never use in our oral discourse. Written signs, however, have an additional semantical feature above and beyond the spoken words they represent in that they denote at the same time spoken sounds (phonemes), spoken syllables, and spoken words, and hence, carry a double level of abstraction: a written sign denotes a spoken word, which in turn denotes a concept from the real world. In logographic systems such as hieroglyphics, visual signs denote whole words. In syllabic systems, the visual signs represent syllables, and in phonetic alphabets, the letters represent phonemes. Words in the latter two systems are represented respectively by some combination of either syllables or letters, depending on the system.

It is at the syntactical level that written language begins to more radically distinguish itself from speech. Punctuation serves a function beyond semantically reproducing the natural pauses and reflections of speech; mainly, it provides a syntactical structure to language that is quite different from that of speech. Writing encourages a formal structuring of language consisting of sentences, paragraphs, sections, and chapters largely absent in spoken language. Analysis of the transcription of most oral discourse reveals that spoken language is not generally organized into grammatically correct sentences. In fact, the very term *grammar* betrays its association with writing through its etymology. The Greek term for "letter" (as in letters of the alphabet) is *gramma*. There are similar associations in other languages — *grammates* in the Latvian language means book. In short, grammar was not formalized before writing, just as there was no such thing as spelling before writing, and no uniform spelling before the printing press.

Grammars are not exact sets of rules. "Were a language ever completely 'grammatical' it would be a perfect engine of conceptual expression. Unfortunately, or rather luckily, no language is tyrannically consistent. All grammars leak" (Stubbs 1980, 39). Before writing, rigorous grammatical conventions were not necessary because any ambiguities could be resolved either by tone, facial expression, or dialogue. Listeners could always ask for an explanation if they did not understand what was said. This is not the case with writing, where readers are basically on their own.

In addition to the formal distinctions between written and spoken language, there is also the empirical change of semantics and syntax that the storage features of writing encouraged. Within the semantic domain, one observes a marked increase in abstract words and terminology as a language acquires a written form. A comparison between the lexicography of Homer (as found in the transcriptions of his orally composed poems) and that of the ancient Greek philosophers and playwrights of the fifth and fourth centuries B.C. reveals the development of a new written vocabulary. The new lexicon of written words is rich in abstract terminology appearing in the language for the first time, and old words take on additional new abstract meanings (Havelock 1963).

Within the syntactical domain, the major structural change with writing is the appearance of the prose system, which incorporates many more analytical features than the oral tradition. With writing, not only do visual syntactical elements appear, such as the spaces between words and the separations of sentences, paragraphs, and chapters, but the creation of permanent records through writing also gave rise to new syntactical elements such as charts, tabulations, tables of contents, and indexes.

In the language of mathematics, the semantical domain or lexicon consists primarily of precisely defined notations for numbers such as 0, 1, 2, 100, 1/2, 0.4, and square root of 2; mathematical operations such as +, -, x, and ÷; and logical relationships such as >, <, and =. The other semantical elements unique to mathematics are its definitions and axioms, such as those found in geometry, number theory, and other logical systems. The language of mathematics differs from natural language such as spoken English in that the semantic relationship between the signal — in this case, a visual sign like a numeral — and the phenomenon being represented as an abstract number is totally unambiguous. In written languages, there are often ambiguities between the written signal and the spoken word being denoted. With the exception of totally phonetic languages such as Spanish and Finnish, there are also ambiguities of pronunciation. George Bernard Shaw's famous example (Stubbs 1980, 51) of using "ghoti" to render the spoken word "fish" — taking "gh" from *enough*, "o" from *women*, and "ti" from *nation* — dramatically demonstrates the pronunciation ambiguities of the English alphabet.

The precision of the semantical conventions of mathematics also extends to the syntactical domain. The basic syntax of the language of mathematics is that of logic. Mathematical syntax, unlike that of spoken or written language, is totally unambiguous. The rules of grammar that govern speech and writing are subject to conflicting interpretations while those of mathematics are not. The language of mathematics also introduces unique syntactical structures not found in natural languages, such as proofs, theorems, and lemmas.

The language of science includes the semantical elements of speech, writing, and mathematics, but it also introduces new semantical units unique unto itself. These include quantitative concepts like mass, force, velocity, mole; qualitative concepts like organic/inorganic, animate/inanimate, solid/liquid/gas, and intelligence; and theoretical concepts like inertia, entropy, valence, and natural selection. As is the case with the language of mathematics, the semantics of science is characterized by precise and unambiguous definitions even though much of the terminology that is employed corresponds to words that appear in everyday spoken language. In spoken English, mass can refer to either volume or weight. In physics, mass is precisely defined in terms of weight, but a careful distinction is made between mass and weight. An object's mass is universal, but its weight depends upon what planet or in what gravitational field it finds itself. A traveler in outer space might experience weightlessness but never masslessness.

As was the case with semantics, the syntax of science includes the structure of speech, writing, and mathematics. Science also introduces its own syntactical elements, however. The three most important elements, the ones which in a sense define the nature of science, are: (1) the scientific method; (2) the classification of information or data (taxonomy); and (3) the organization of knowledge such as the grouping of scientific laws to form a scientific theory. The centrality of the classificatory and organizational structures is due to the fact that science is defined as organized knowledge. The scientific method, with its elements of observation, generalization, hypothesizing, experimental testing, and verification, is the key element which defines the character of science. It is the scientific method that qualifies science as a distinctive language rather than a carefully organized scholarly activity like history, which also makes use of organizational principles and other modes of language, namely, speech, writing, and mathematics.

The language of computing includes all of the semantical and syntactical elements of the other four modes of language. It also possesses its own semantical and syntactical elements by virtue of the activities of both its programmers and its end users. The semantics of the programming languages and end-user software programs specify computer inputs and outputs. The syntactical structures of programming languages and end users' software formalize the procedures for transforming inputs into outputs. These syntactical structures are basically unambiguous algorithms for ensuring the accuracy and the reliability of the computer's output. The syntactical structures that arise in a programming language or a relational database differ from the other language modes so that the user can take advantage of the computer's rapid information-processing speeds.

An Evolutionary Chain of Languages

The primary mode from which all the other modes of language derive is speech. Each of the other modes incorporates elements from its predecessors. Without speech, there would have been no writing and no mathematical notation; without writing and mathematics, there would have been no science. Without science, there would have been no computing.

Each mode of language incorporates the features of the previous modes. Computing incorporates the features of all the previous modes: speech, writing, mathematics, and science. Science incorporates speech, writing, and mathematics. Writing and mathematics arose at the same moment in history. They therefore only incorporated the features of speech, albeit different ones. It is worth tracing the history of the development of these five modes of language and detailing the history of human communication and information processing. The story begins with speech or the spoken word.

SPEECH

Speech, or spoken language, is the earliest and most basic form of human language. It is essential to all information processing and, perhaps, even to human thought itself. There are two schools of thought, one headed by Edward Sapir (Sapir 1921) and Benjamin Whorf (Whorf 1964) and the other by Lev Vygotsky (Vygotsky 1962), which claim that a spoken language determines the thought processes of its users and that thought arises from speech, not the other way around. Without fully subscribing to their theories, this analysis is, nevertheless, based on the premise that spoken language plays an essential role in all human information processing. Spoken language provides the categories with which we express our thoughts and ideas and organize our information.

Speech is most often regarded solely as an instrument of human communication. This is its primary function. The first uses of spoken language were for the purposes of communication. As this type of language evolved, however, its users discovered that verbal information could be recorded and stored through the mechanisms of poetry and song. This discovery gave rise to the oral tradition among preliterate people whereby vital information necessary for their survival, their identity, and their sense of history was recorded or stored in their tales, epics, and legends. Oral language was used as a device for processing, storing, retrieving, and organizing information. The preservation or storage of information and knowledge in preliterate societies was achieved through the memorization of folktales or myths. These tales or legends were not merely entertaining stories told in an impromptu manner. They were, in fact, very carefully organized to provide listeners with the basic information required in their society. In his *Preface to Plato*, Eric Havelock refers to the storyteller as a "tribal encyclopedia." The information in the story was also organized to facilitate its memorization as a way of preserving or storing it. As noted earlier, the use of rhyme, rhythm, meter, alliteration, plot, and other "literary" devices were used to aid the storyteller's memory. These techniques of information storage and organization, however, imposed a set of constraints on oral society, one which discouraged innovation and personal expression (Havelock 1963).

WRITING[1] AND MATHEMATICS

The origin of writing can be traced to the Sumerian culture at the end of the fourth millennium B.C. The writing systems that originated in Egypt, China, the Indus Valley, and Meso-America are independent, and their invention followed a different set of circumstances than those in Sumer.

Arguments have been made that the idea of writing was transmitted from Sumer to Egypt and China, but this is speculative. This discussion focuses on the invention of writing in Meso-

[1] Elements of this section are based in part on a 1983 working paper of the author's entitled "Cross Cognitive Impacts of the Notations for Writing and Numbers." Although the working paper has not been published, it was cited by Denise Schmandt-Besserat in "Before Numerals" (*Visible Language* 18, winter 1984, 48–60). The author benefited immensely from extended discussions and correspondence with Dr. Schmandt-Besserat and wishes to acknowledge her inspiration for the preparation of this section. To her I accord credit for whatever is of value here.

potamia since the Sumerians were not influenced by any other writing systems and their innovation grew out of the earlier notational schemes of tallies and clay accounting tokens.

Ironically, writing did not evolve from storytelling or the need to preserve oral tales and legends, but out of the need to enumerate and to make records of commercial transactions. Written records began with quantitative rather than qualitative data. "Writing was not a deliberate invention, but the incidental by-product of a strong sense of private property" (Speises, quoted in Meyers 1960). The transmission of the idea of writing to Egypt and China is assumed to have taken place along trade routes, and this reinforces the notion that commercial, rather than cultural, activities stimulated the development of writing. This should not be surprising, since quantitative information is far more difficult to remember than verbal information. It is much easier to remember a story with a narrative and various characters or personalities than a set of numbers.

TALLIES

The first form of human notation, tallies, were not written but were made either by making marks on an object — such as etching notches on a stick or a bone — or by collecting counters such as pebbles, grains, twigs, or shells. Tallies are based upon the principle of one-to-one correspondence in which each item being enumerated is matched by a mark or a counter depending on the system (Schmandt-Besserat 1984a, 48–60). The number of identical marks or counters used for the tally was equal to the number of items being enumerated, such as the number of sheep in a flock, the number of animals killed in a hunt, or the number of full moons since the last meeting of two clans. The major limitation of tallies is that they record only the quantitative feature of the set of objects being enumerated, leaving the identification of the objects to context or to the memory of the enumerator. Tallies are, therefore, qualitatively nonspecific. They can satisfy only simple, low-technology cultures where only a few obvious items of daily life require accounting. They cannot be used, for example, to tally a complex set of commercial products.

The oldest form of tallies (and, hence, of human notation) currently identified are notched animal bones and antlers dating to about 15,000 B.C. in the Middle East (Marshack 1964, 743; Schmandt-Besserat 1987, 44–48). These tallies show a series of notches or incised strokes in variable numbers, which can be interpreted as standing for collections of unknown items in a one-to-one correspondence. Marshack has proposed that some of the European Paleolithic tallies served to keep calendrical notations. Whatever the meaning of the Middle Eastern notched bones, it is clear that quantitative or numerical notation preceded written notation.

As primitive as they were, tallies produced an impact on the human mind at the cognitive level. They represent the first use of visual symbols to store, manipulate, and display information. In turn, this fostered abstract thought processes as items of daily life came to be dealt with through the means of an abstract notation. Three deer, for instance, could be represented by three abstract marks. Tallies segmented reality. As a substitute for the deer, the marks isolated the deer from their context and abstracted them for scrutiny. Tallies also encouraged objectivity by separating the information from the knower, and at the same time displaying that information for all to use. Tallies replaced verbal communication with objective nonverbal data removed and abstracted from their original context. Tallies created a new reality and set the stage for counting, mathematics, and accounting. As far as we know, they were the first artifacts to designate quantitative data nonverbally. Because the counters could be repeated infinitely, they increased the amount of data that could be handled by humans. Tallies provided a tool for organizing information and laid the foundation for the future development of other information-processing techniques.

ACCOUNTING TOKENS

The next step in the development of notational technologies were clay accounting tokens. Denise Schmandt-Besserat's exhaustive study of tokens revealed that these clay artifacts, two to three centimeters long, were used for the enumeration and accounting of agricultural staples. Their use began in the prehistoric Middle East, circa 8000 B.C., at the very beginning of the agricultural age, which also had its roots in the Middle East. Each uniquely shaped token designated some measure of an agricultural commodity or product. The system which started with twenty-four kinds of tokens grew to 190 different types of tokens by 3300 B.C., just before the advent of writing and abstract numerals in Sumer. The shapes included spheres, disks, cones, tetrahedrons, biconoids, ovoids, cylinders, and triangles, which were further differentiated with incisions and punched marks. The tokens designated agricultural products and commodities such as: measures of grains (barley, wheat, and emmer); jars of oil; livestock, primarily sheep and goats, differentiated by age, sex, and breed; wool, cloth, and different types of garments; measures of land expressed in terms of the amount of seed required to sow it; and service or labor (Schmandt-Besserat 1978).

Tokens and tallies share many characteristics. This suggests that the token system may have evolved from pebbles. Both tokens and tallies are palpable symbols that are easy to manipulate and are used in one-to-one correspondence with the objects being enumerated. Both fragment reality in that they abstract data from the context in which they are generated and from linguistic discourse and hence are not dependent on any particular spoken language. This helps explain the wide distribution of tokens throughout the ancient Middle East among people who spoke different dialects and languages. This is similar to today's use of Arabic numerals, whose use spans many different cultures and linguistic groups.

Despite their similarities, tokens differed from pebbles and other tallies in very significant ways. Tokens were not randomly collected objects put to the secondary use of counting. They were man-made artifacts prepared specifically for the purpose of accounting. Compared to a series of more or less identical counters such as grains or pebbles, tokens were molded into distinctive shapes that were easy to recognize and simple enough to be systematically reproduced. Compared to unspecific tallies, each token shape stood for a specific commodity. In other words, tokens communicated both quantitative and qualitative information (Schmandt-Besserat 1987).

Tokens functioned as metaphors, translating words for economic concepts into a tactile medium. This was a radical change in data processing. Compared to a group of pebbles that could not be understood out of context, the tokens became a reliable tool for data storage. Their meaning could be understood and translated unambiguously into speech by anyone initiated into the system. The tokens were a tool for data manipulation. Like tallies, they made possible the processing and retrieval of large amounts of data, but tokens had yet another advantage. They also permitted a new dimension, that of diversification, which allowed many different commodities to be dealt with at once. As new commodities requiring enumeration arose, a newly shaped token could be created. The elements of an inventory consisting of different types of commodities could be manipulated simultaneously. As with tallies, the data could be easily laid out and displayed, and hence grasped both visually and physically. Tokens, however, could be organized and reorganized into different piles or more sophisticated patterns according to any number of possible categories, which facilitated and enhanced classification. In a certain sense, the token system is the forerunner of the abacus as well as spreadsheet analysis.

Tokens also had the potential of acting as a means of communication. While tallies could be used only as personal memory aids for items known to the enumerator alone, tokens were universal and systematized. A token came to represent the same product to individuals of entirely different communities speaking different languages and separated by great distances.

The main drawback was that the system became cumbersome. Although it was open and complete and could be expanded at will to convey new meanings, the creation of new shapes and new markings was bound to reach a stage of saturation or overburdening. This created the need for a new form of data processing, which eventually led to the invention of writing and abstract numerals.

THE COGNITIVE IMPACT OF THE TOKEN SYSTEM

The token system changed the way people stored, manipulated, displayed, and communicated information and, in particular, the way people counted. Tokens were a new tool for counting and computing. The unique feature of the counters was that they specifically designated the object being enumerated by the coding of qualitative information through a three-dimensional shape. Each token with its unique shape was used to enumerate one, and only one, object or commodity and no other. For example, ovoids could count only jars of oil. Tokens, in other words, merged the representation of the qualitative and quantitative features of a commodity. There was no token representing the abstract unit of "one." There were only tokens representing one unit or one measure of a specific commodity. When counting with tokens, it was not possible to separate the nature of the item counted from that of the quantity. Abstract counting was not possible with the token system; tokens did not represent abstract numbers.

There exists, in fact, no evidence for the use of abstract numbers at the time tokens first appeared, nor during the time of their major use and development from 8000 B.C. to 3100 B.C. Schmandt-Besserat's analysis (Schmandt-Besserat 1984a, 1987) of notched bone (Marshack 1964) led her to the conclusion that these tallying devices were used exclusively in a nonmathematical, one-to-one correspondence mode. There does not seem to be the grouping of notches into sets corresponding to cardinal numbers such as 2, 3, 5, or 10. Analysis of the use of tokens has yielded a similar result, namely, that tokens were used to enumerate in one-to-one correspondence like tallies with the added feature that their distinctive shape qualitatively designated the specific commodity being enumerated. Evidence for the way in which tokens were used comes from groups of tokens representing specific transactions found stored and bundled in clay envelopes. The analysis of these assemblages of tokens indicates that each enumeration of a product was tied to multiple repetitive occurrences of the tokens designating that product. That is, the tokens were used in one-to-one correspondence: *n* ovoids were used to designate *n* jars of oil. On the other hand, *n* jars of oil are never represented as a set of *n* tokens, each representing the abstract unit "one" tied to the ovoid token for a jar of oil. The syntax of the tokens is such that they never functioned as adjectives modifying a noun. The syntactical structure of an abstract number, such as "three," acting as an adjective modifying a noun (or a noun phrase) such as a "jar of oil" in the phrase "three jars of oil" does not appear before writing.

Given the lack of evidence for the use of abstract numbers with tallies and tokens, Schmandt-Besserat (Schmandt-Besserat 1984a, 1987) postulated that tokens reflect a stage in the evolution of counting, before abstract numbers were used, known as concrete counting. Many historians of mathematics view the evolution of counting as having taken place in three main stages:

1. One-to-one correspondence with no precise concepts of numbers.
2. Concrete counting with the use of special numerical expressions for counting different specific classes of items.
3. Abstract counting with abstract numbers which are universally applicable.

Schmandt-Besserat's analysis (Schmandt-Besserat 1984a, 1987) of the archaeological data supports the claims of historians of mathematics and suggests that the use of tallies represented counting in a one-to-one correspondence with no specific concept of number. Tallies permitted people to count larger quantities of items than the vocabulary of enumeration that was available to them in their spoken language. Tokens acted as concrete numbers. As we have seen, they had meaning only as units of the commodity they designated and enumerated, and not as abstract absolute quantities. Concrete numbers, such as "a brace" of partridge or "a yoke" of oxen, like tokens, cannot be used to designate "two" as an abstract number and then be used to enumerate other objects. A brace of sandals is meaningless; instead one must refer to them as a "pair of sandals," that is, as a concrete number or as "two sandals" where "two" operates as an abstract number. The invention of tokens made it possible for enumeration to become specific and to develop "concrete counting."

The dominant technologies of a culture, whether they are informational, mechanical, or economic, are interrelated and follow a similar pattern of evolution, reinforcing each other until a new level of technological innovation or breakthrough arises and new social patterns emerge.

The history of information processing in the Middle East during the period 15,000–3000 B.C. is one of major breakthroughs followed by long periods of stability. The breakthroughs — tallies, tokens, and writing — may be viewed as "break boundaries" between radically different economic and social systems. The tally system was in use for thousands of years by hunting and gathering societies. It remained static, not undergoing any significant morphological or functional changes as long as hunting and gathering was the economic base of the culture. It is not a coincidence that the token system, a major elaboration and significant modification of the tally system and a breakthrough in information processing, emerges around 8000 B.C., just as agriculture is emerging as the economic base for a new form of society.

The token system became a stable technology of information

processing for the next 5,000 years until another major shift in economic life occurred, namely, the urbanization of Sumer at Uruk between 3350 B.C. and 3100 B.C. (Kramer 1956). With urbanization, tokens began to evolve new forms and applications. Clay tokens were used to record the payment of tributes and taxes to those who maintained and administered the irrigation system upon which all agriculture depended. Tokens became an administrative tool that allowed political and economic control of the agricultural countryside by urban centers of commerce and trade. Cognitive breakthroughs and advances associated with these innovations resulted in a chain reaction in which new information-processing techniques arose, including abstract numeration, logographic writing, and phonetic coding.

The new advances in the technology of information, such as tallies, tokens, abstract numerals, and writing, did not occur in isolation. Rather, new information forms emerged as changes took place in the social and economic environment. The cognitive foundations of the previous level of information technology served as a starting point for satisfying the new information demands of a changed environment. For example, the tally system was transformed into the token system by the demands for new accounting procedures required by the new agricultural lifestyle. Similarly, urban life brought with it pressures for a more complex accounting system as the number of items to be enumerated and the number of social interactions to be taken into account dramatically increased. These demands manifested themselves in a spurt of creative activity in which the simple token system was articulated and expanded, and new structures were added, which eventually led to the invention of writing and abstract numbers. An overview of this complex process of innovation, described below, can be obtained from Table 1, which shows the stages through which the use of tokens passed and the cognitive process associated with them.

COMPLEX TOKENS, ENVELOPES, AND IMPRESSED LOGOGRAMS

Changes in the way quantitative information was processed can be attributed to changes in the socioeconomic system of its users. The token system during the long purely agricultural period from 8000 to 3350 B.C. remained remarkably stable with the appearance of a limited number of plain or simple tokens corresponding to basic agricultural commodities of the time, such as wheat and sheep. Between 3350 B.C. and 3100 B.C., the token system underwent a metamorphosis which can be attributed to the pressures of urbanization. With the rise of urban life and, in particular, the monumental temple complexes such as Eanna at Uruk, the largest economic institution of its time, there suddenly appears in the ruins, along with "other administrative material such as seals and sealings" (Schmandt-Besserat 1988a), complex tokens representing a number of finished manufactured products, no doubt produced in the urban workshops of Uruk and other population centers. The great majority of tokens from this period are found in the temple precincts, indicating that they were associated with a coercive taxation and redistribution system of a strong central government. This interpretation of the token findings is reinforced by Sumerian reliefs of the period showing citizens delivering their goods to the temple (Kramer 1956).

Another important development in the token system occurs with the appearance of clay envelopes which functioned as containers to hold a number of plain tokens. It is presumed that the envelopes were used to consolidate all of the tokens employed in a single transaction. The surfaces of the envelopes were marked with the seals of the individuals involved in the transaction, and hence the envelope served as a receipt or a contract (Schmandt-Besserat 1984b, 47). The envelopes had the disadvantage, however, of masking their contents because of the opacity of their clay surfaces. This problem was overcome by impressing the tokens to be contained in the envelope onto the wet surface of the envelope.

Breakpoints in the History of Information Processing with Tallies and Tokens

TABLE 1

TIME FRAME	MORPHOLOGY	ASSOCIATED COGNITIVE ADVANCE OR NEW COMMUNICATION CAPABILITY
15000–8000 B.C.	TALLIES Notched sticks, etched bones, pebbles, shells	ENUMERATION: ONE-TO-ONE CORRESPONDENCE
8000–3400 B.C.	SIMPLE TOKENS Three-dimensional clay objects	CONCRETE COUNTING AND NUMBERS Recordkeeping and data storage Information or data processing Primitive sign and symbol capability Classification — control — centralization
3400–3250 B.C. (Urbanization at Uruk)	COMPLEX TOKENS Greater variety of three-dimensional clay objects • Multiplication of shapes • Incisions and punctuations • Perforations	SIGN AND SYMBOL ARTICULATION — METHAPHOR Abstracting classes through common shapes but differentiating subjects Complex classification Complex data processing and recordkeeping Specialization

3250–3150 B.C.	**Token impressed envelopes** Clay containers of tokens whose outside surface have been impressed with the tokens that are contained inside.	**Abstracting 3-dim. artifacts into 2-dim. signs** Primitive set theory or chunking of data More abstract symbolization — visual metaphors Translation of tactile into visual
3150–3100 B.C.	**Impressed tablets** Two-dimensional clay tablets upon which token forms have been impressed.	**Greater abstract symbolization** The token is no longer physically present Ability to work in a totally two-dimensional medium
3100 B.C.	**Incised tablets** Two-dimensional clay tablets upon which: • signs have been incised which replicate the shapes of tokens, • ban and bariga token have been impressed, • new incised signs have been added.	**Advent of writing and abstract numerals** Logographic writing Abstract numerals Phonetic writing

This technique revealed the contents without the necessity of breaking the envelope and it also had the unexpected and certainly unplanned side effect of creating the first two-dimensional logograms (Schmandt-Besserat 1987).

A logogram is a two-dimensional visual sign abstractly representing the written form of a single word. Egyptian hieroglyphics and Chinese characters are examples of logograms. Two different types of logograms can be distinguished: pictograms, which represent the words they designate pictorially; and ideograms, which represent the ideas of the words they designate. A logogram can also simply represent the word symbolically. The system of writing that makes use of logograms is called logographic writing.

The purpose of the token impression was to reveal the content of an envelope, not to substitute token impressions for tokens. But this is precisely what happened. Approximately fifty years later it was realized that the actual tokens contained within an envelope were totally redundant and their use was dropped. The curved envelopes quickly evolved into flat two-dimensional tablets with the impressions of tokens upon their upper surfaces. The evolution of envelopes into tablets represents a major breakthrough in information processing. The flat tablet became the prototype of all two-dimensional writing surfaces, including the pages of a book and the CRT video display of a computer.

It is important to note that the impressed logogram, semantically, was no different from a three-dimensional token. The logograms were two-dimensional negative imprints of the three-dimensional tokens. They were immediately recognizable to the users of the token system because they shared the same outline and bore the same markings as the actual tokens. In fact, the first tablets are not true writing but merely the permanent records of an accounting system based on tokens in which the tokens themselves were actually discarded once their imprints had been made. From the point of view of function or informational content, impressed logograms on clay tablets operated exactly the same way as the three-dimensional token system. The medium changed, however, and the way that information was displayed changed radically from a collection of palpable three-dimensional clay objects to abstract symbols arrayed in two-dimensional patterns on the flat surface of a clay tablet. The shift from three-dimensional artifacts to logograms in the form of negative images impressed on two-dimensional surfaces triggered a chain reaction that resulted in the invention of logographic writing, phonetic coding, and abstract numerals.

PREPARING THE WAY FOR NUMERALS AND WRITING

The manufacture and manipulation of tokens and impressed signs promoted the development of hand-eye coordination and fine motor skills essential for writing. More importantly, tallies, and especially tokens, introduced the visual bias that characterizes written notation. Tokens stressed the importance of uniformity. They were the first class of artifacts whose uniformity of shape was key to their function. Tokens were used as repeatable symbols in much the same way that numerals and the letters of the alphabet are. The visual bias created by the use of uniform and repeatable tokens increased when the tokens were used to imprint the outer surfaces of clay envelopes and, later, the impressed tablets. The pattern of token markings arranged in neatly spaced lines within a two-dimensional array permitted the user (or reader) to encompass the whole field of data in a single glance.

Tallies, tokens, and impressed signs also paved the way to writing and abstract numerals on the cognitive level. Each of these methods of record keeping can be regarded as phases in the development of abstract notation. Impressed token markings increased the level of abstraction. The two-dimensional format of the impressed tablets was more abstract than the three-dimensional tokens in that the negative impression of the tokens could not be grasped in the hand and manipulated like three-dimensional tokens. The loss of volume, concreteness, and tactility removed the symbols one step further from the physical three-dimensional reality for which they stood. The display of information within a two-dimensional array, however, permitted the user to see the information in a new light. By being able to observe more data at a single glance, the user of the tablet system began to think more globally about the data. More abstract patterns of classifying and analyzing the data became possible. If the medium is the message, then the message of the two-dimensional display of information on tablets was that of abstraction, classification, analysis, uniformity, repeatability, and the power of the visual.

It is hypothesized that the development of tallies, tokens, impressed signs, numerals, and writing were phases in the development of metaphorical (visual) thinking and that the earlier notational forms created a perceptual and cognitive environment conducive to the invention of writing. To summarize how tallies, tokens, and impressed signs led to numerals and writing, let me iterate the many features which abstract numbers and writing shared with the previous record-keeping devices (tallies, tokens, impressed envelopes, and impressed tablets).

Each system employed metaphors and symbols and hence abstracted information by fragmenting reality from its representation in a sign system, thus creating an artificial medium for the expression and notation of ideas in which the basic elements were easily manipulated. Each system consisted of uniform elements, which were used in a repeatable fashion within the context of a linear two-dimensional array (truer of later stages of the token system), and thus produced a visual bias that required good hand-eye coordination for their creation and use. The application of these systems promoted objectivity, and hence standardized uniform measures and a system of economic and political control.

THE ADVENT OF WRITING AND ABSTRACT NUMERALS

Once tokens were used to produce two-dimensional impressed signs, a series of rapid but fundamental changes took place that transformed the token-based notational system for "concrete numbers." There emerged two notational systems, one for abstract numerals and one for written words. Abstract numerals were built out of the two impressed token signs for the small and large measure of wheat, the "ban" which came to represent the number 1 and the "bariga" which represented 10. The second notational system was a full-fledged writing system for transcribing the words of spoken language into visual signs incised on the clay surface of a tablet with the sharp end of a stylus. The shapes of the original impressed token signs were imitated in the new incised format and formed the first vocabulary or lexicon of the new written language. Words other than those for commercial commodities emerged based on the principle of representing words pictographically or ideographically. Paradoxically, the very first incised signs did not represent the words pictorially, but represented the shape that the accounting token corresponding to that word made when impressed into a clay tablet. Writing began as an extremely abstract symbol system.

The transformation that took place circa 3100 B.C. was far more than a change of form from impressed tokens to incised signs. It was a change of concept. Instead of signs that fused qualitative and quantitative information, there were two separate types of notation: one verbal and one numerical. Incised signs were used to transcribe spoken words or qualitative information — that is, rendering spoken language in a written form. The two former impressed logograms, the ban and the bariga, were used to express abstract numbers. With the emergence of two notational forms, a qualitative one with written words and a quantitative one with abstract numerals, the language of the written word and the language of mathematics emerged as two separate and distinct systems for recording information. The transformation that led from impressed logograms to incised pictographic writing and abstract numerals involved the following five major changes:

1. The transition from token impressions to incised signs: The two-dimensional impressions of tokens onto clay tablets inspired the emergence of a new form of signs. The new incised signs took the form of small designs traced with the sharp end of a reed stylus on the surface of a clay tablet while the clay was still wet. The designs mimicked the shape and form of the token impressions. The stylus was pressed lightly onto the clay surface to produce the number of short strokes necessary to draw a sign. Circa 3100 B.C., the practice of imprinting the clay tokens was dropped completely, with the exception of the two signs for the two grain measures which continued to be impressed.

How did this technique for creating signs arise? One possible motivation for the use of incised signs rather than impressed tokens is that the former technique allowed the creation of new signs for words that had no corresponding token. Still another reason is that there could be no confusion with the old notational system. The sign for a ban of wheat repeated twice followed by a jar of oil sign can only be read as two jars of oil if the jar of oil sign is incised. If the jar of oil sign had been impressed, however, the three signs could possibly have been misinterpreted as a tablet from the old system and read as two bans of wheat and a jar of oil. By using incised signs for the words and impressed signs for the numbers, the new tablets could never be confused with the older tablets consisting of impressed signs only.

The transition between impressed and incised signs was rapid, a matter of only fifty years, or two generations. The commodities signs that had been produced by impressing the corresponding token into the clay were now rendered with a stylus by sketching the token's outline and repeating its markings. Each of the incised logograms in this new written language became a visual metaphor or a pictographic symbol of the former tokens. The critical empirical evidence supporting this hypothesis for the invention of writing is the existence of a one-to-one morphological correspondence between the first pictographs representing the most usual Sumerian commodities, such as grain, livestock, oil, and textiles, and the three-dimensional tokens used to represent these goods before writing.

2. An increase in the repertory of signs — pictography and ideography: The representation of tokens by an incised sign on the tablets led to the idea that any concept could be conveyed by an incised sign. The use of the stylus, in other words, led to the invention of new pictographic and ideographic signs completely unrelated to the tokens and designating new concepts unrelated to agricultural commodities.

3. New semantic and syntactic values of the symbol system: Not only did the incised signs differ from the impressed logograms morphologically, but also semantically and syntactically. The impressed logogram for a sheep carried the meaning of "one sheep" and functioned as a modified noun (that is, a concrete number) in which both quality and number were denoted. The incised logogram was, on the other hand, a more abstract symbol which denoted an unmodified noun and hence stood for a "sheep." A separate symbol was required to denote the number of sheep even if it was only one.

4. Signs to express sounds — phonetic writing: Pictography was unable to satisfy one particular need of the Sumerian scribes: it could not render an individual's name. It was beyond the capability of pictograms or ideograms to represent each individual by a miniature portrait. Sumerian ingenuity provided the solution. Names were expressed phonetically. Logograms, used to write an individual's name, were read phonetically as the sound of the word they represented in Sumerian. The sign for oil was read as the sound "i"(long e) and the sign for arrow was read "ti." When associated together in rebus fashion as the Sumerians would have done it, the two signs spell "E.T." — the name of the cinema hero from outer space.

5. Signs to express plurality — numerals: The first written numerals representing abstract numbers appear at exactly the same time as incised logographic writing (Schmandt-Besserat 1984a, 56). The actual signs to represent numerals were made by impressing different combinations of the cone and sphere tokens representing, respectively, the small and large measure of wheat, the ban, and the bariga. These numerals still retained the visual metaphor of the one-to-one correspondence with a pile of pebbles. The number 3 was represented by the impression of three cones (or three wedges) and the number 4 by four cones (or four wedges). The new element of abstraction in notation was the use of a new sign, the circle, that is, the impression of the sphere, for the number 10. This created an economy of notation since 43, for example, could be represented as four circles and three wedges instead of forty-three wedges. This system was more efficient than the older system of tallying. It was an early and primitive predecessor of the system of Roman numerals, which would represent 23, for instance, as two Xs and three Is or XXIII. The two impressed logograms used to represent the abstract numbers 1 and 10 still retained their function as "concrete numbers" for grain. It was left to the scribe to decide, according to the context, whether the wedge was the numeral 1 or a ban of grain, and whether a circle was the numeral 10, or a bariga of grain.

THE BIFURCATIONS OF IMPRESSED LOGOGRAMS

As we have seen, both logographic writing and the notation for abstract numbers arose 5,000 years ago from clay accounting tokens, which in turn first made their appearance 10,000 years ago. The token system underwent a linear sequence of development and enrichment from 8000 B.C. to 3100 B.C. which comprise the following steps or stages:

1. plain tokens;
2. complex tokens;

3. tokens in clay envelopes;
4. impressed logographs on clay envelopes containing tokens; and
5. impressed logographs on clay tablets.

None of these changes represents a bifurcation, but they do represent either an increase in the variety and number of tokens or a morphological change of the tokens, that is, a change in their physical form. The meaning or function of a three-dimensional token circa 8000 B.C., or an impressed logogram created by pressing a token into wet clay circa 3100 B.C., were the same. During the above five steps, neither writing nor abstract numerals emerged. The next development in notational technology was profound and followed shortly after the introduction of the two-dimensional clay tablet. The impressed logograms on clay tablets not only underwent semantic and morphological variations but were subject to a series of Prigogenian bifurcations, discussed below.

The two logograms for the small and large measure of grain, the ban and the bariga, remained unchanged morphologically. They continued to serve as "concrete counters" for the two measures of grain, but they also took on a second semantic function: they came to represent the abstract numbers 1 and 10, respectively. At the same time that this transition took place, incised logograms also came into being. All of the existing impressed logograms other than the ban and the bariga logograms ceased to function as "concrete counters"; they changed both morphologically and semantically. Each impressed logogram — with the exception of the two grain signs — was replaced by incised signs that graphically resembled the older impressed logograms they were replacing. In addition to this morphological change, the signs also changed semantically. They no longer functioned as "concrete numbers" but as logograms designating single spoken words or phrases. For example, the impressed sign that had designated "one sheep" was replaced by an incised sign that was read simply as "sheep." In addition to the transformation of the old impressed signs into incised signs, other incised signs began to appear, designating concepts which had not been part of the original token vocabulary consisting of approximately 190 agricultural commodities and manufactured goods.

Not long after their introduction some fifty years later, incised signs took on a second semantic function. As well as ideographically or pictographically coding a single word, they were also used to phonetically code those phonemes and words that had the same pronunciation as the word they represented ideographically or pictographically. The incised sign for an arrow, ti, also represented its homonym, life, also pronounced "ti."

The sudden explosion of semantic function in 3100 B.C. marks the beginning of literacy and numeracy. To understand the interplay of the cognitive elements that contributed to this development, we require a more sophisticated model. The work of Iya Prigogene and his theories of bifurcation and social amplification illuminate this process which took place in Sumer 5,000 years ago.

The emergence of the complex notational structures of the numerals and writing that suddenly appear in 3100 B.C. can be described as a series of bifurcations involving the two-dimensional "concrete counter" logograms whose impression was made by pressing the original three-dimensional tokens from which they were derived into a wet clay tablet. The emergence of abstract numerals and writing can be described in terms of the following three separate bifurcations:

Bifurcation 1: the split of two-dimensional impressed logographic signs into two morphologically and semantically distinct sets of signs — the incised signs and the impressed signs in which incised signs are used exclusively for verbal notation or writing and impressed signs used exclusively for accounting and numerical notation;

Bifurcation 2: the split of the impressed logograms for the ban and the bariga into two semantic functions — the old one as "concrete numbers" for small and large measures of grain and the new one as the "abstract numbers" for 1 and 10, respectively; and

Bifurcation 3: the split of the semantic function of the incised signs, in which they are used not only to code spoken words ideographically or pictographically, but also phonetically.

It is impossible from the archaeological data (Schmandt-Besserat 1979, 27–40; Schmandt-Besserat 1988b, 1–175) to give the exact dates of the above bifurcations. Logic dictates that Bifurcation 3 must have occurred sometime after Bifurcation 1 since the incised signs would have had to have been established as logograms designating a word with a particular sound before they could have been used to phonetically code the homonyms of that word. Empirical evidence supports this conclusion since fewer examples of the phonetic use of the incised signs are found compared with the more concrete use of the signs as logograms.

The relative timing of Bifurcations 1 and 2 can be determined by a process of elimination and a linguistic analysis of the archaeological data. Since it is only the relative timing of Bifurcations 1 and 2 that is in doubt, there are only three logical possibilities as illustrated in Table 2, namely, Bifurcation 1 occurs: (i) before; (ii) at the same time as; or (iii) after Bifurcation 2.

The fact that impressed logograms functioning as "abstract numbers" do not appear before the existence of incised signs and when they do appear are always used to modify an incised sign, indicates that Bifurcation 2 did not occur before Bifurcation 1. There are no cases where the ban and the bariga signs are used to modify an impressed token logogram such as the jar of oil or the sheep sign.

The absence of any tablets in which incised signs are repeated many times to function as "concrete numbers" in the way tokens or impressed-token logograms functioned indicates, on the other hand, that Bifurcation 1 did not occur before Bifurcation 2. This leaves only one possible conclusion: Bifurcations 1 and 2 occurred at precisely the same time, and hence, model B in Table 2 best explains the data. It would seem that the emergence of abstract numerals and ideographic writing were tied to each other and arose at the same point in history.

Each of the three bifurcations represents the birth of a new form of semantic or numerical expression in which a new abstract branch splits off from an older, more concrete one.

Bifurcation 1 represents the emergence of true writing using an ideographic or pictographic coding scheme in which each word is uniquely represented by a visual sign, that is, the written form of the word. Bifurcation 1 also involved the splitting of abstract logograms, each designating a single word, from impressed logograms or "concrete numbers" that designated both a quantity and a quality. The incised logogram abstracts away from the impressed logogram the sense of number, leaving a purely qualitative system of notation; hence, the beginning of true writing, which is capable of coding a spoken language into written signs.

Bifurcation 2 represents the emergence of a notation for "abstract numbers," which eventually evolved into our present system of Hindu–Arabic numerals. It also involved the splitting of "abstract numbers" from "concrete numbers," and hence the creation of a purely "quantitative notation" in which the quantity being notated is totally independent of the nature (or quality) of the objects being enumerated.

Bifurcation 3 took place within the confines of verbal writing and represents the separation of logographic and phonetic coding. These two systems of coding, which emerged within fifty years of the beginning of true writing, also exhausted all of the known possible techniques for coding spoken language into written signs. (Exceptions such as the Morse code or the electronic binary coding

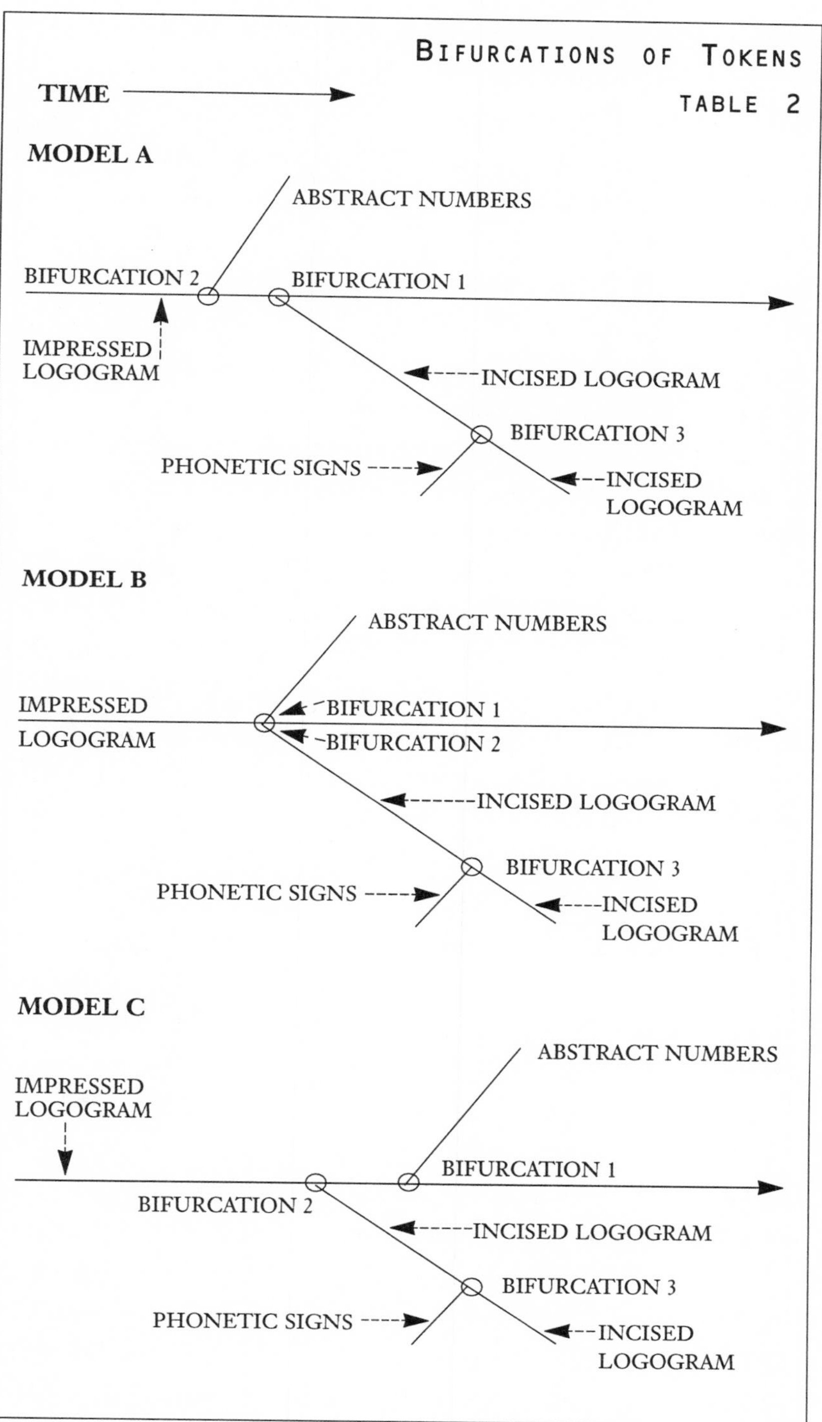
BIFURCATIONS OF TOKENS
TABLE 2
TIME
MODEL A
ABSTRACT NUMBERS
BIFURCATION 2
BIFURCATION 1
IMPRESSED
LOGOGRAM
INCISED LOGOGRAM
BIFURCATION 3
PHONETIC SIGNS
INCISED
LOGOGRAM
MODEL B
ABSTRACT NUMBERS
IMPRESSED
LOGOGRAM
BIFURCATION 1
BIFURCATION 2
INCISED LOGOGRAM
BIFURCATION 3
PHONETIC SIGNS
INCISED
LOGOGRAM
MODEL C
ABSTRACT NUMBERS
IMPRESSED
LOGOGRAM
BIFURCATION 1
BIFURCATION 2
INCISED LOGOGRAM
BIFURCATION 3
PHONETIC SIGNS
INCISED
LOGOGRAM

of information with computers represent a form of coding in which written language and not spoken language is coded.)

Bifurcations 1 and 2, which occurred at the same moment in history, represent the original split between the quantitative and qualitative modes of notation and analysis that exists today. Bifurcations 1 and 2 reversed the 5,000-year-old trend that had been established with tokens of combining quantitative and qualitative description and analysis. The separation permitted a specialization of notations, and therefore, techniques of analysis that led to many other advances. This split also had an impact at the subliminal level and set the pattern of information processing. It contributed to a cognitive style of analysis in which the quantitative and qualitative are carefully distinguished and are treated separately. The Western characteristic of specialization had its roots in Bifurcations 1 and 2, and thus began some 5,000 years ago.

In addition to specialization, the three bifurcations led to other Western characteristics of abstract thought. Western thought patterns compared with those of the East are abstract and deductive rather than concrete and analogical. As shown, each of the bifurcations involves a transition from a less abstract to a more abstract notational element (see Table 3):

Bifurcation 1: from token impressions to ideographic writing;
Bifurcation 2: from "concrete numbers" to "abstract numbers"; and
Bifurcation 3: from ideographic coding to phonetic coding.

The "concrete counters," whether in the form of tokens or impressed logograms, coded information by analogy: quantitatively, by one-to-one correspondence, and qualitatively, by pictographic symbols. Abstract numbers and phonetic signs code information using logical algorithms, and promoted the abstract logical thought of Western civilization, which was the heir to the token system.

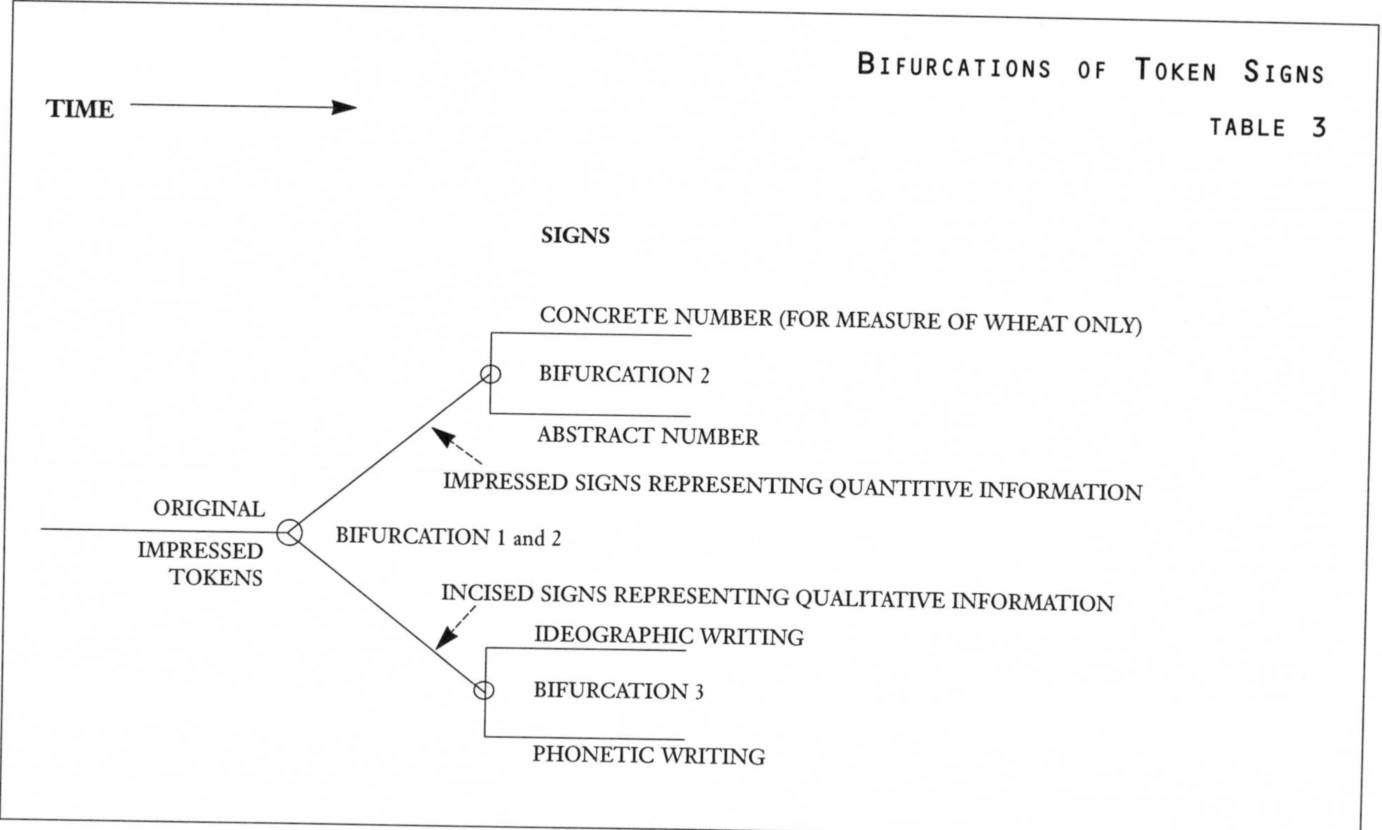
BIFURCATIONS OF TOKEN SIGNS
TABLE 3
TIME
SIGNS
CONCRETE NUMBER (FOR MEASURE OF WHEAT ONLY)
BIFURCATION 2
ABSTRACT NUMBER
IMPRESSED SIGNS REPRESENTING QUANTITIVE INFORMATION
ORIGINAL
IMPRESSED
TOKENS
BIFURCATION 1 and 2
INCISED SIGNS REPRESENTING QUALITATIVE INFORMATION
IDEOGRAPHIC WRITING
BIFURCATION 3
PHONETIC WRITING

With this in mind, one can date and place the origin of Western thought to circa 3100 B.C. when, in Sumer, abstract numerals, ideographic writing, and phonetic coding emerged, setting the pattern for all subsequent intellectual developments that were derived from these systems of notation (McLuhan and Logan 1977, 373–83; Logan 1986a).

THE EVOLUTION OF PHONETIC WRITING SYSTEMS

Phonetic Syllabaries

At first, only names and monosyllabic words were rendered phonetically. Later, phonetic signs for certain commonly used syllables developed so that polysyllabic words could be rendered phonetically, and a set of syllabic signs came into existence. The Sumerian writing system eventually evolved into a mixture of ideograms and phonetic syllabic signs. When the Sumerians, who spoke a non-Semitic language, were conquered by the Semitic-speaking Akkadians, circa 2300 B.C., the conquerors used the Sumerian writing system to transcribe their own spoken language. The ideographic signs retained the same meaning but were pronounced according to Akkadian speech. The phonetic signs, however, retained their Sumerian pronunciation and became more abstract because their sound was meaningless to Akkadian speakers. The Akkadian writing system shifted to a much greater extent to phoneticization, until eventually there was a finite, but complete, set of phonetic syllabic signs capable of rendering all the spoken sounds of the Akkadian language. The logographic signs, with a few exceptions, began to disappear. Eventually, the writing system was reformed to a system of only sixty syllabic signs. The reform of the writing system during the reign of Hammurabi, circa 1750 B.C., was accompanied by reforms of the system of weights and measures and by the establishment of a legal code throughout the Hammurabic empire. The reformed syllabic writing system and the other accompanying reforms represented another breakpoint in information processing. They were brought about by the pressures for the uniformity and regularity required to govern an empire consisting of many different city-states and ethnic groups.

The Phonetic Alphabet

The next stage in the evolution of notational systems was the invention of the phonetic alphabet. The phonetic principles that emerged in the transition from impressed to incised logograms and that were employed to create syllabaries were used to code the individual phonemes of a spoken language. This development took place in two stages. In stage one, the Canaanites (or Proto-Phoenicians) developed a writing system with only twenty-two signs, one for each of the consonants of their spoken language. The absence of a system for coding the vowels did not present major difficulties because in Semitic languages the consonants determine the root word. The function of the interior vowels is to provide grammatical information such as tense (as is the case with the English words *sang*, *sing*, *song*, and *sung*).

The actual invention of the alphabet by the Canaanites is thought to have resulted from a borrowing of a subsection of the Egyptian writing system in which a set of twenty-two uniconsonantal signs were used to phonetically render proper names. The use of the Egyptian uniconsonantal signs paralleled the first use of phonetic signs by the Sumerians. The sound value of the letters of the Canaanite alphabet (or the Egyptian signs) is the same sound as the initial consonant of the spoken word that the sign designates pictographically. For example, the second letter of the Semitic alphabet, "beit," which means house, was represented originally by a picture of a house, □, and had the sound value of the letter B.

The Addition of Vowels

The second stage in the development of a fully phonetic alphabet occurred when the Greeks borrowed the Phoenician alphabet and transformed it to suit their language. They added three new consonants to accommodate the special sounds of their language. Their major breakthrough, however, was to create signs for the vowels by transforming the aspirant consonants of the Phoenician alphabet, not required for Greek pronunciation, into the vowels of the Greek language. For example, "h" became an "e." Without the vowels, it would have been impossible to render Greek or any other Indo-European language with an alphabet consisting solely of consonants. Imagine trying to distinguish "idea" from "duo" without vowels, or the following group of words from each other: bat, bait, beat, bet, boat, bout, bought, and but.

With the addition of vowels, the Semitic alphabet was transformed by the Greeks into the totally phonetic alphabet capable of translating every spoken word of the Greek language into a unique combination of letters and every combination of letters into a unique pronunciation. No system of writing has ever surpassed the totally phonetic alphabet for its capacity to represent any spoken language unambiguously with a minimum number of visual signs. Subsequently, the phonetic alphabet was borrowed and transformed by many different cultures and became the writing system for hundreds of languages, particularly those of Europe, western Asia, and northern Africa, but not necessarily limited to those areas.

WRITING AND THE ALPHABET EFFECT: A NEW MODE OF INFORMATION PROCESSING

Writing goes beyond the mere transcription of spoken language. Although writing is a medium whose content is spoken language, its uses differ from those of speech. It organizes and stores information quite differently than speech, and in effect is a different form of language. Written language has evolved in ways quite different from speech. The rules for the construction of speech and prose are quite different. Prose is not recorded speech but a much more formal organization of information. The closest form of writing to speech is poetry, which is meant to be heard. Prose is not meant to be heard and, in fact, the most efficient form of accessing the information contained in prose writing is to read it silently.

The epic poems of Homer provide a unique opportunity to compare the oral and the written forms of language. The poems were composed and presented orally long before they were transcribed in their present form "sometime between 700 and 550 B.C." (Havelock 1978, 3). Before their transcription, there was no particular order in which the various episodes were presented by the bard who performed them. Once the verses were committed to writing, however, they became objectified, an artifact that could be studied and scanned visually, their components compared, and then edited into some temporal order, as is the case with Homer's verses. "The story pieces are sorted out and numbered so as to achieve the effect of a single overall time sequence which moves forward but with interruptions, flashbacks, and digression to an appointed end. Thus arose the arrangement of our present text . . . this arrangement being the work of the eye, not the ear, a work achievable only when the various portions of the soundtrack had been alphabetized" (Havelock 1978, 19).

The very act of transcribing an oral composition requires an ordering. The text becomes a physical artifact that can be "looked at, reflected upon, modified, looked at repeatedly, shown to others, etc" (Havelock 1978). The medium helps determine the mode of organization: "Due to the transitory character of the acoustic medium, spoken language is organized by continuity, connectivity, and integration . . . in addition to intonation. By contrast, because of the visual medium, written language is organized by discreteness, and segmentation" (Ludwig 1983, 39).

The objectification of information, or the separation of the knower from the knowledge that writing permitted, encouraged abstraction, systematization, and the objectivity of scientific thought. Phonetic writing, particularly alphabetic writing, encouraged classification and codification of information. Alphabetizing provided a natural way of ordering verbal information. It is interesting to note that the order of the letters of the alphabet never changed despite the fact that it was passed from one culture to another and adopted by so many different languages. The names and shapes of the letters changed but not the order of their presentation when the alphabet is recited as "abcdef."

The use of the alphabet as a writing code promoted and encouraged the skills of analysis, coding, decoding, and classification. Each spoken word to be transcribed had to be broken down into its phonemic components, each of which was then represented by a unique letter of the alphabet. The impact of information processing due to alphabetic writing was enormous because it introduced a new level of abstraction, analysis, and classification into Western thinking. At the same time that the Hebrews adopted alphabetic writing, they codified their law in the form of the Ten Commandments and introduced, for the first time in the history of humankind, monotheism and the notion of a prime mover or first cause.

The Greeks also made enormous intellectual strides shortly after their adoption of the Phoenician alphabet and its modification to include vowels. They were the first culture to develop deductive logic, abstract science, and rational philosophy. The Hebrews and the Greeks were also the first societies to write fairly objective histories of their own nations.

While none of these intellectual developments can be causally linked, Marshall McLuhan and I postulated (McLuhan and Logan 1977) that the use of phonetic writing systems created an environment conducive to the development of codified law, monotheism, abstract science, deductive logic, objective history, and individualism. All of these cultural innovations occurred primarily between 2000 B.C. and 500 B.C. in the closely linked cultures found between the Tigris-Euphrates river system and the Aegean Sea. The introduction of the phonetic alphabet represents a break boundary because of the tremendous intellectual and cultural fallout that followed from its use (Logan 1986a).

ZERO AND THE PLACE NUMBER SYSTEM

One of the themes of this chapter has been the way in which one notational system influences the development of another through a process of evolution. We have also seen that although the notation systems for writing and abstract numerals are independent, they arose together from impressed token logograms. Without abstract numerals, there might not have been writing, and without writing, there might not have been abstract numerals. The two systems of notation share a number of features. Writing employs a basic set of visual signs (for example, a set of logograms, a syllabary, or an alphabet) to transcribe or record the sounds of a spoken language into an array of visual signs. The language of mathematics also notate numbers and mathematical operations with a set of visual signs. For example, the operation of addition is represented with the + sign. At first, writing and mathematical notation were used to store the verbal or mathematical utterances of spoken language. As their use expanded, however, literature and mathematics emerged as languages in themselves with their own peculiar techniques for processing, retrieving, and organizing information quite distinct from that of speech.

The cross-pollination between the two systems of notation continued as writing evolved its more sophisticated phonetic elements. The alphabet influenced quantitative analysis because it stimulated an analytic and rational approach to the organization of qualitative information. This led to the development of abstract science and deductive logic by the ancient Greeks. In turn, logic and science stimulated the need for exact and precise quantitative analysis and measurement. This led to axiomatic geometry, the elements of Euclid, the quantitative astronomy of Ptolemy and Aristarchus, the Pythagorean obsession with numbers, the mechanics of Archimedes, and the botanical and biological classification schemes of Aristotle.

The Semitic alphabet stimulated quantitative analysis directly by becoming the basis of a number system in which each letter was assigned a numerical value. The first nine letters represented 1 through 9, and the next nine letters represented 10, 20, . . . through 90. There were also letters for 100 and 1,000. Any number between 1 and 100 could be represented by a maximum of two letters. For example, 18 would be written as חי where (ח) is 8 and (י) is 10. The Semitic number system was not a place number system in that 81 was not written (יח) but rather as (עא) where (א) is 1 and (ע) is 80. The system was a forerunner of the place number system in that it contained one of the essential features for such a system; namely, that all numbers between 1 and 9 had their own unique ideographic signs. The element lacking from the alphabetic number system that prevented it from serving as a place number system was the concept of zero and the zero symbol.

The alphabetic number system (which the Greeks and Hindus also employed with their alphabets) also had all the elements for a place number system except the concept of zero. The place number system and the concept of zero were inventions of Hindu mathematicians as early as 200 B.C. The Hindu writing system at the time was alphabetic, as was their number system. Once the Hindu mathematicians developed the notion of zero, or *sunya*, as they called it, they quickly devised a place number system.

Sunya means "leave a place" in Sanskrit and indicates that the zero or sunya concept arose from recording abacus calculations. If the results of a calculation was 503, this could not be written as "5""3" because it would be read as either 53 or 530, but if instead the result was written as "5" "leave a place" "3", the number being designated would be interpreted properly as 5 hundreds, no tens, and 3 units. "Leave a place" or sunya soon evolved into the abstract number of zero, 0.

The Arabs used the Hindu system and transmitted it to Europe, where it arrived in the fifteenth century. The Arabs had translated sunya or "leave a place" into the Arabic *sifr* or cipher, the name we still use for zero as well as the name for the whole place number system itself. Our present-day term *zero* derives from the shortened version of the Latin term for cipher, *zepharino*. The place number system brought with it many advances in mathematics, including simple algorithms for arithmetic, negative numbers, algebra, the concept of the infinite and the infinitesimal, and hence, calculus.

One of the mysteries associated with the invention of the place number system is why the Greeks, the inventors of vowels, who made such great advances in geometry and logic, did not discover zero. The explanation lies in the Greeks' overly strict adherence to logic, which led Parmenides to the conclusion that non-being (and hence, nothing) could not "be" because it was a logical contradiction. The Hindus, on the other hand, had no such inhibition about non-zero. In fact, they were positively inclined to the concept of non-being since it constituted their notion of nirvana (Logan 1986a; Logan 1979, 16).

THE LANGUAGE OF MATHEMATICS

Numeric notation, like writing, grew out of the system for recording the payment of tributes with clay tokens. The language of mathematics grew out of spoken language and the need for a mathematical notation for enumeration. The mathematical statements expressed in the numerals and the signs for mathematical operations can always be translated into spoken language or writing. For example, the equation $1 + 2 = 3$ may be spoken or written as, "One plus two is three." The language of mathematics may be regarded as a medium whose content is the mathematical concepts of spoken language, namely, abstract numbers and mathematical operations, such as addition, subtraction, multiplication, and division. Once a mathematical language or notation emerged, however, it took on an existence of its own and evolved in ways quite different from either speech or writing. This is due to the unique methods of processing information that its use stimulated. A mathematics notation allowed abstract mathematical operations to emerge that could never have been carried out in a person's head. The notation became a tool of investigation and invention. It suggested ways of generalizing results, and thus led to new concepts.

The concept of and notation for zero is a perfectly good example of this. In addition to the place number system, the concept of zero, originally notated as a dot and then later as a small circle, resulted in a number of new mathematical ideas. Not long after introducing zero, the Hindu mathematicians invented negative numbers which they notated by placing a dot above a number (Logan 1986a; Logan 1979). For example, $\overset{\circ}{3}$ represented minus three or three below zero, which the notation literally suggests since the 3 sits under the zero sign. The zero sign, sunya — "leave a place" — was also used to represent the unknown. This allowed the development of algebra because the mathematician was able to "leave a place" for the unknown and deal with it as though it were also a number. Here, we see clearly how the existence of a notation (for zero, in this case) allowed the ideas of negative numbers and algebra to develop. The notation for zero caused mathematicians to think of zero as a number like 1 or 2, and hence to consider mathematical operations with this number such as addition, subtraction, multiplication, and division. The addition or subtraction of zero leads to no change and multiplication by zero yields zero, but the division of a number by zero leads to an interesting result, namely, infinity, which was another new concept or invention of the Hindu mathematicians. The idea of infinity led them also to the notion of the infinitesimal, a concept that was an essential element in the development of calculus (Logan 1986a; Logan 1979).

THE IMPACT OF QUANTITATIVE AND QUALITATIVE NOTATION AND ANALYSIS

There is no way short of speculation to determine how the idea for numerals and writing arose from the impressed token logograms. Because of the difficulty of establishing causal relations among cognitive processes, even when historical data are readily available, it is difficult to determine how qualitative and quantitative notational systems influenced each other's development. It is evident, however, that the new level of abstraction that pictographic writing established created an environment conducive to the development of abstract numbers and vice versa. It is clear from Schmandt-Besserat's work that the processes of literacy and numeracy are more closely linked than is commonly acknowledged. Numbers and phonetic writing emerged from a single progenitor, the clay token, in which both quantitative and qualitative information had been merged. The separation of this information into two distinct streams of written words on the one hand, and numerals on the other, opened new avenues of abstract thought and information processing. Ideographic and phonetic writing represented the first means of supplementing and extending verbal language. Numerals representing abstract numbers made possible new techniques of quantitative analysis. Both these developments mark the beginning of objective learning or the separation of the knower from the knowledge (Schmandt-Besserat 1985, 149–54).

The quantitative (from the Latin *quantus* — how great) and the qualitative (from the Latin *qualis* — of what kind) are two key categories of Western thinking that have, from the philosophical thoughts of the ancient Greeks to contemporary social science, been regarded as distinct and independent modes of analysis. The common origin of quantitative and qualitative notation from clay tokens argues against the notion that these two categories form a dichotomy. The fact that the two forms of notation emerged at the same point in history indicates the cognitive power that the interplay between the quantitative and qualitative can release.

Other breakthroughs in information processing can be associated with this interplay. The letters of the alphabet were used to notate abstract numerals until the Hindus, under the influence of alphabetic literacy, invented zero and developed the cipher system (Logan 1986a; Logan 1979). It is not an accident that zero and the place number system were the invention of mathematicians who used the alphabet. The place number system and the alphabet share a number of features that contribute to their abstract nature:

1. Each system contains a small number of elements: twenty-six letters (for the English alphabet) and ten numerals.
2. They form a complete set so that the total set of possible spoken words can be represented alphabetically, or any number, no matter how large or small, can be expressed in terms of some combination of the ten numerals.
3. The individual elements of the two systems, the letters and the numerals, are atomic — that is, they are identical and repeatable.
4. The values (sound or numerical) of the aggregate elements (words or numbers) of the system depend not only on their atomistic components (the letters or numerals of which they are composed) but also on their order or syntax. In other words, both the letters and their order determine a word, and the numerals and their order determine a number. For example, ON is not the same as NO, nor is 18 the same as 81.

These similarities in the two systems illustrate two points: the alphabet was probably a stimulus to the development of the place number system; and quantitative and qualitative notational schemes are not all that different and require many of the same fundamental cognitive skills.

The common origin and emergence of quantitative and qualitative notation is only one of the indicators of the overlap of these two categories. An abstract number denotes the quantity of objects in a set, but it can also describe a quality of the set. In addition to their numerical values, the abstract numbers 1, 2, or 3 also denote the qualities of oneness, twoness, or threeness. For example, in the saying "Two is company and three is a crowd," the numbers 2 and 3 are abstract numbers and yet also describe qualities. The relationships described are quite independent of the actual 2 or 3 individuals who constitute the "company" or the "crowd."

Abstract numbers themselves possess both quantitative and qualitative features. The difference between the abstract numbers 3 and 4 is quantitative in the sense that 4 is one more than 3, but the difference is not purely quantitative; it is also qualitative in that 3 is a prime number and 4 is not, but rather a perfect square. Another overlap of the two categories of letters and numerals is the way in which their use is combined for the purposes of creating classification schemes; for example, the call numbers for library books and license plates where letters and numbers are used together.

The link or overlap we have established both empirically and theoretically between the qualitative and the quantitative is confirmed etymologically when the terms used for the profession of writing or numerical accounting are examined. In ancient Babylonian or Hebrew, the term for one who writes and one who counts is identical, namely *spr* (Demsky 1972). There is a similar overlap of meaning of these words in English. A "teller" is one who counts or tells, and the expression "to give an account" can mean either to provide a narrative or an enumeration depending on the context in which the word is used. One finds a similar overlap in other languages. In German, the word *zahlen* has a similar double meaning denoting either counting or telling.

Numeracy, the ability to conduct mathematical calculations, is a cognitive skill that emerged as a result of the development of quantitative notation. Although there are individuals who can do calculations in their head without using notated figures, the origin of these calculational techniques required the existence of numerals. While literacy and numeracy are quite distinct cognitive skills, it is obvious that there is a strong overlap between them. The claim that some individuals are gifted with words but not numbers or vice versa has little basis in our understanding of the origins of these skills. Marked differences in an individual's literate and numerate skills could be due to a disparity in interest rather than intrinsic abilities. Historical research has indicated that literate and numerate skills are related and associated with each other. Split-brain research supports this hypothesis, as both the literate and the numerate activities seem to be concentrated in the left hemisphere of the brain.

The importance of these observations for contemporary education is not obvious. They do indicate, however, that current instruction in primary school of reading and writing, on the one hand, and mathematics on the other, requires greater integration. Drawing parallels between the two notational systems could certainly help students to understand the abstract nature of the alphabet and the place number system. It might help those who are strong in math but weak in reading, or vice versa, to use their strengths with one notational system to better understand the other. These suggestions are purely speculative but certainly worthy of further examination and research. The use of the computer in education might provide exactly the correct environment for such an integration as the computer treats all abstract notation in more or less the same manner. In fact, the computer may be regarded as a device for the manipulation of abstract symbols, whether they are alphabetic or numerical.

THE ADVENT OF ABSTRACT SCIENCE

Abstract science is the systematic organization of knowledge that the languages of writing and mathematics gave rise to. It is a language in its own right because it possesses unique semantical and syntactical elements including its special vocabulary, analytic tools, classification schemes, and employment of the scientific method.

The fact that abstract science with its special way of organizing information arose in an intensely literate and numerate environment is not an accident. Writing provided a number of the basic tools of classification and analysis required for scientific thinking. It also provided an environment conducive to innovation. "A technological society can only exist where there is a fair amount of literacy. For oral habits of mind emphasize memory and training . . . [and this] suggests an inability to deal with new and hitherto inexperienced situations. Given the flexibility and scope of the written word . . . it is not surprising to find that the human mind is freed to deal with events and situations in a more innovative and complex manner" (Hershbell 1978, 90). Just as context and plot aid oral memory, the categories, methodology, theories, and generalizations of science play a similar role. They provide a context or a schema that aids the memory and allows an individual to consider more data with a single sweep of the mind. Just as the cognitive structures due to writing "impose an organization that can be used to guide the retrieval process" (Anderson 1980, 218), the cognitive structures of science perform a similar role.

Further evidence for the link between mathematics, writing, and science is that the vanguard of Sumerian science were the teachers in the scribal schools. "Within its [the school's] walls flourished the scholar-scientist, the man who studied whatever theological, zoological, mineralogical, geographical, mathematical, grammatical, and linguistic knowledge was current in his day, and who in some cases added to the knowledge" (Kramer 1959, 2).

Both writing and the existence of a system of mathematical notation were essential ingredients in the development of abstract science. But science as an organized body of knowledge, and a methodology, goes far beyond writing and mathematics as a unique system for processing, storing, retrieving, organizing, and communicating information. Science is not just knowledge about nature but rather knowledge organized in a special way so that results can be generalized and new information can emerge or be easily collected from the existing knowledge base. The language of science is constructed in such a way that it is conducive to classification. For example, the categories that Aristotle created were the beginning of scientific language. "The theories of categories is partly a theory of language and partly a theory about reality . . .

Aristotle did not think of the structure of language as mirroring the structure of reality. But he did believe that there are specific items of language and reality the consideration of which forms the crucial link between the two" (Moravcsik 1967).

Because science organizes information in a unique way, we consider it as a language quite distinct from speech, writing, and mathematics. If science is regarded as a medium, then writing and mathematical notation are its content, as are its observational data, categories, hypotheses, and theories. Writing and mathematical notation play a special role in that they are the media through which science is reported. Spoken language is also a part of science in that it facilitates communication among scientists.

Developing an understanding of nature and applying that knowledge to practical affairs has been a universal feature of all societies since the beginning of time. There developed within the framework of Western civilization, beginning with the ancient Babylonians, Egyptians, and Greeks, a more systematic and abstract approach to scientific inquiry, which we shall label abstract or theoretical science. Abstract science is characterized by the use of deductive logic, mathematics, a systematic organization of information using appropriate classification schemes, and a methodology for formulating and testing hypotheses — the scientific method.

Understanding the historical roots of abstract science is important for a number of reasons: first, abstract science represents a way in which both the qualitative and quantitative modes of analysis merge through science's use of the languages of speech, writing, and mathematics to study and describe nature; secondly, abstract science has developed into a language itself because of the unique way that it processes, stores, retrieves, organizes, and communicates information; and thirdly, abstract science provided one of the essential ingredients for the development of still another language, computing.

The roots of abstract science may be traced to the origins of literacy and numeracy in Mesopotamia. The existence of a writing system led naturally to a more formal system of classification. The origins of classification can be traced to the compilation of lists of plants, fish, rivers, city-states, and rulers that were prepared by the teachers in Sumerian scribal schools to teach reading and writing. The emergence of codified law and phonetic writing in Mesopotamia also contributed to a spirit of classification that later influenced the development of Greek science. The Mesopotamian notation for numbers and the need for quantitative analysis that commerce inspired led to the development of arithmetic, an essential ingredient for abstract science. Another essential element in the origin of science was Egyptian geometry. Because of the annual flooding of the Nile River, and the need to measure land area before and after the flood, the Egyptians made a number of empirical discoveries that formed the foundation of geometry. The first Greek scientists, beginning with Thales, formalized Egyptian geometry by proving a number of empirically derived theorems from a set of first principles or axioms. They extended and generalized the Egyptian results and so developed not only axiomatic geometry but deductive logic, another essential ingredient of abstract science.

Deductive logic led to the notion of causality and a "first mover." The pre-Socratic philosophers attempted to explain all natural phenomena in terms of first principles. Thales, for example, believed all things to be derived from water. For Anaximander, the first principle was the neutral substance Apieron from which opposites emerged. Aniximenes described all of nature in terms of air, while Heraclitus used fire as his first principle. The ancient Greek philosophers might have derived the idea of a unifying principle guiding the universe from the Hebrews' notion of one God, Jahweh, the creator of the universe. The Hebrews developed a notion of causation and the prime cause which they attributed to Jahweh.

> In fact, the name Jahweh incorporates the notion of causing to come into being. "The enigmatic formula in

> Ex. 3:14, which in the Biblical Hebrew means 'I am what I am,' if transposed into the form in the third person required by the causative Jahweh, can only become 'Jahweh asher jihweh (later jihyeh), 'He Causes to Be What Comes into Existence.'" The Hebrew notion of causation did not develop along the logical scientific lines of the Greeks. Rather, they incorporated it into their unique sense of history, with its promise of the future, the Promised Land, and their role as the Chosen People. (Albright 1957, 87)

The phonetic alphabet with which the Greeks transcribed their spoken language provided them with a model for the abstraction, analysis, and classification essential for formulating abstract science. The written word, as transformed by the printing press, also played a key role in the development of modern science as a result of the superior storage and organization of information that print made possible.

The most convincing argument for the importance of writing and numeracy to the origin of science, however, is the simple empirical evidence that all breakthroughs in abstract science have occurred in literate and numerate cultures. As a result of their careful observations, many preliterate societies developed a good understanding of many aspects of nature, such as the medicinal properties of plants or the behavior and habits of animals. What distinguishes their knowledge of nature from abstract science is the way in which they organized their information. The difference between abstract science and primitive science is that the former leads to new discoveries because of the systematic way in which it organizes information. The organizational structures of abstract science naturally pose questions to the scientist which stimulate the process of discovery. Abstract scientists will go out of their way to perform experiments to test the universality of their organizational structures, whereas preliterate cultures are content to describe nature as they encounter it. They also limit their studies of nature to that which is immediately practical to them. The ancient Greeks saw little practical consequence in their first studies of amber, but they pursued their investigations for their own sake, and what they discovered eventually contributed to our present understanding of electricity.

Science as a mode of language and hence as a system of information processing, has led to a series of breakthroughs in our understanding of nature on a large number of fronts, including astronomy, physics, chemistry, biology, and medicine. It has also had a major impact in social fields through the social sciences and the field of information and computer science. Science as a language has been influential in practical endeavors through the various fields of engineering it has spawned and through the Industrial Revolution for which modern science acted as a model or paradigm.

Thomas Kuhn explains how the information-processing techniques of science are propagated through the articulation of its paradigms (Kuhn 1972). Kuhn's model of science and how it changes can be used to show the transition from one regime of information processing to another. When speech could no longer handle the information-processing needs of agricultural society, visual notational systems were introduced that led to abstract numerals and ideographic writing. This paradigm was repeatedly articulated, which, in the domain of writing, led to phonetic coding, syllabaries, and finally to the phonetic alphabet. In the field of mathematics, the notation of abstract numbers led to signs representing arithmetic operations, zero, the place number system, algebraic notation, vectors, tensors, and calculus. When the information-processing capabilities of mathematics and writing were no longer sufficient, abstract science with its methodology of generalizing, hypothesizing, and testing arose to fill the gap.

Finally, when science outgrew its information-processing capabilities, computers were developed, bringing contemporary society to its present level of information processing. The paradigm of the mainframe computer as an automated number-cruncher has already undergone a number of articulations resulting in the minicomputer, the microcomputer, the notebook, the hand calculator, and a host of software applications among which word processing, spreadsheets, databases, accounting packages, multimedia, and desktop publishing are the most popular.

THE ADVENT OF COMPUTING

The computer may be regarded simply as a device for the automated processing, storing, retrieving, organizing, and communicating of information. It basically manipulates abstract symbols, whether they are words, numbers, equations, or databases. Computers change inputs into outputs in a systematic, reliable, and relatively rapid manner. The operations of the computer are closer to its French name, *ordinateur*, or "that which orders," than its English name, which better describes its humble origins as an automatic calculator. To understand its implications, however, it is useful to define computing as a language in the way speech, literature, mathematics, and science are defined as language. The language of computing may be considered to include all of the information resident in computers, as well as all of the information and techniques needed to operate computers. The information resident in computers may be regarded as the content of the language of computing, and includes spoken language, written language (literature), mathematics, and science. The techniques for using computers may be seen as the grammar of the language of computing. The reason that computing may be considered as another level of language is that its strategies for processing and communicating information are very different from that of the other modes.

Just as literacy and numeracy were essential for the development of abstract science, so in turn was modern science essential for the development of computers. Modern science provided the necessary technical ingredients such as electronics, circuitry, the magnetic storage of information, mathematics, logic, Boolean algebra, solid state physics, and telemetry. Science also provided the motivation to design computers because progress in a number of fields required computational techniques for calculating complex mathematical problems which could not be solved using standard numerical techniques. The computer, with its ability to perform large numbers of simple calculations at high speed, provided the solution to these problems. Complex calculations could be broken down into a large number of smaller and simpler problems using the techniques of numerical analysis. The solutions to the simpler problems could be calculated by developing algorithmic procedures which in turn could be automated. This is the basic principle behind the computer-based solutions of the complex mathematical problems that arise in the domain of science, as well as governmental administration (for example, collecting taxes and tabulating census results) and big business (for example, maintaining airline schedules or financial records).

Language: Cognitive, Technical, and Social Interplay

Upon those that step twice into the same river
different waters flow.
(Heraclitus)

The evolution of the five languages suggests a model or theory for the development of communication and information-processing systems, based on the idea that all innovations have a cognitive, social, and technological component. They are the three basic dimensions of the process of cultural change.

Our notion of technology should not be limited to hardware inventions, but rather, as the Greek word *tekhne*, meaning "skill" or "art," indicates, should incorporate all human tools. These include the physical tools used to organize the material world, the conceptual and cognitive tools used to organize information, and socioeconomic tools or institutions used to structure or organize society. Examples of physical tools include pottery, the plow, and the steam engine. Examples of conceptual tools include spoken language, the alphabet, the number system, scientific laws, and mathematical algorithms. Socioeconomic tools or institutions include tribal organization, priesthoods, the family, markets, and bureaucracies. But not every technological tool belongs exclusively to one class. Some tools like the clay tablet, the printing press, and the computer have both a physical and a conceptual component. This division is tacit in the terminology which distinguishes between the hardware and software components of a computer. The school is another example of a mixed system: the bricks and mortar are physical, the curriculum is conceptual, and the administration is socioeconomic.

Every technological tool (whether physical, conceptual, socioeconomic, or mixed) involves new ways of organizing information and hence requires new cognitive skills. And vice versa — every cognitive breakthrough manifests itself in terms of one or more new tools or organizing principles whether they be the airplane, Einstein's Theory of Relativity, or the Magna Carta.

The five languages which have been under discussion may be considered as basically conceptual technological tools, but each one requires one or more physical artifacts. The development of speech required the evolution of a biological artifact, namely, the Homo sapiens speech apparatus consisting of an enlarged pharynx, a specialized larynx, greater muscular control over the tongue, and the use of the pharynx, the mouth, and the nasal cavity as sound resonators. With all the other forms of language, the physical artifacts are man-made tools. For writing and mathematics, the first artifacts were simply a clay tablet and a wooden stylus. With the passage of time, other physical artifacts were used, including parchment, papyrus, paper, ink, pens, lead, pencils, the compass, the ruler, the protractor, the printing press, and the typewriter. At first, the physical artifacts of abstract science were the same as those for writing and mathematics. But as science became increasingly more empirical and experimental, the physical tools of science came to include the instruments of experimentation and observation. Examples include the astrolabe, the telescope, the pendulum, the inclined plane, the microscope, the test tube, the petrie dish, the Faraday disk, nuclear reactors, and particle accelerators.

The physical artifacts of computing were at first mechanical calculators, which were eventually replaced by electric and electronic components such as electro-mechanical relays, vacuum tubes, transistors, integrated circuit silicon chips, magnetic memory devices such as tape drives and disk drives, monitors, and printers. The conceptual components of the language of computing are the operating systems, the software programs, and all of the theoretical ideas that led to computers, including Charles Babbage's original model of a mechanical computer circa 1835.

In tracing the development of the five modes of language, we have noted how each new development, including the very origin of speech, resulted from the interplay of the cognitive or conceptual tools, physical technology and socioeconomic factors at work in the culture. It is believed that speech developed in order to provide a level of control over and coordination of the tools that erect toolmaking proto-*Homo sapiens* began to manufacture and implement. "Philip Lieberman (Lieberman 1975) has claimed that there are similarities between the cognitive patterns — roughly, the mental maps or plans — which underlie the chipping of stone tools of the sort made during the Paleolithic period, and the syntactic structure of human language" (Hickerson 1980, 25).

Cognitive tools and physical technology are two resources at the disposal of human innovators, and the needs or demands of society are often the motivating force. Necessity is the mother of invention, yet invention does not occur in a vacuum. All of the previous innovations in a culture provide the resources, both cognitive and physical, for the next level of innovation. The previous innovations also contribute to changes within the socioeconomic system which give rise to new social demands. Each new invention, technological innovation, or discovery gives rise to new technical capabilities, new cognitive abilities, and new social conditions. These then interact with the existing economic, political, social, cultural, technical, and cognitive realities of the culture to set the stage for the next round of innovation. Thus, technological change in our model is part of an ongoing iterative process. It began with the inception of *Homo sapiens*, the toolmaker, and continues to this day at an ever-quickening pace.

The history of ideas has the appearance of a linear evolution of technical change. Periods of technological stasis are punctuated by moments of violent technological change and breakthroughs. This gives rise to a picture of the "great moments" in the history of technology in which "great individuals" charged with genius make earth-shattering breakthroughs. These innovations are slowly digested by the rest of society until the next genius comes along to once again "shake things up" and start a new process of change.

In fact, between explosions of creative energy, a ferment of activity takes place, quietly at the cognitive level of individuals and within the dynamics of the socioeconomic interactions of society. Change is at work not only within these two spheres but also at the interface, where a subtle tension between the two arises, from which change flows. As Heraclitus explained, during a period of great technological change in Greece when the impact of alphabetic literacy was first being felt, "All things happen by strife and necessity." It is the cross-impact back and forth among the technical, cognitive, and socioeconomic spheres that gives rise to technological innovation.

We have seen this model at work in the developments which have gone from the tallies of hunters and gatherers, to the token system of Neolithic Middle Eastern culture, to the emergence of writing and abstract numerals with the urbanization of Sumerian society at Uruk circa 3100 B.C. The Middle Eastern token system grew out of the economic and social needs to enumerate commodities and keep a record of transactions. The need for a notational scheme arose because quantitative information is more difficult to retain in one's memory than verbal information. We know that the storytellers of preliterate societies were able to commit elaborate stories to memory, but there is no evidence to support the retention of quantitative data.

The use of the token system in ancient Sumer allowed administrators to control the flow of wealth and to build a centrally controlled communal irrigation system. This led to greater wealth, the rise of urban centers, the specialization of skills, the proliferation of the variety of goods, and the need for the capability to handle more complex transactions involving a greater variety of commodities and merchandise. This chain of events gave rise to complex tokens, clay envelopes, impressed logograms, and finally abstract numbers and writing. Each development in the chain of events accelerated the rate of discovery and increased the variety of information the system could store, organize, and process. Phonetic writing allowed still more sophisticated information processing, and hence greater control of society by central authorities. The government that arose made use of the new tool of writing to establish legal codes and the increased mathematical abilities to establish standard weights and measures. The centralized government also established and subsidized the scribal schools needed to train government clerks, and hence supported the scholarly activities and research of the teachers, which laid the foundations for scientific thought.

The developments leading to the invention of the alphabet by the Proto-Canaanites, its transfer to Greece by the Phoenicians, and the addition to it of vowels were a result of the commerce that took place among the Egyptians, the Semitic tribes of the Levant, and the Greeks. Each of the breakthroughs in information storage and processing, such as the invention of numerals, writing, tablets, syllabaries, the alphabet, and papyrus, permitted the storage of more and more data and information. This resulted in the need for even more sophisticated information-processing techniques. The adoption and adaptation by the ancient Greeks of the phonetic alphabet led to the advent of abstract science, rational philosophy, deductive logic, axiomatic geometry, objective history, representational democracy, and individualism. While the intellectual ferment was not directly tied to the commercial and administrative activities of the society, a number of links existed: the philosophers who developed new ideas were intimately involved in the practical affairs of state; a number of the pre-Socratic philosophers such as Heraclitus were legislators in their community; Anaxagoras owned an olive press and was a successful merchant; Aristotle was a tutor to Alexander the Great; Archimedes used his mechanical talents to build war engines; the library and museum at Alexandria was state-subsidized and flourished partly because of the successful commercial production of papyrus.

The Roman era, which continued the tradition of Greek learning, was even more oriented towards the practical application of knowledge. The Romans did not advance abstract thinking as much as the Greeks, but they are responsible for the transmission of the fruits of Greek thinking and its application in the world of commerce and civic administration. Roman engineering and architecture were disseminated throughout Europe by their conquering armies. The collapse of the Roman empire resulted in the collapse of literacy and book learning. At the practical level, however, advances in the development of technology continued once order was restored. By the time of the late Middle Ages, mechanics and other practical arts were quite sophisticated.

Modern science emerged in the Renaissance as a marriage of the empirical spirit fostered by mechanics and the return to Greek scholarship and an interest in abstract science. The growth of modern science, engineering, and industrialism, beginning with the Industrial Revolution, is a perfect illustration of the model in which the interplay of the cognitive, technical, and socioeconomic environments results in a chain of innovations, inventions, and conceptual breakthroughs. The economic growth of this period and the institutions of capitalism, mercantilism, and nationalism gave rise to conditions that encouraged technological and conceptual breakthroughs. The spirit of uniformity and specialization which infused the industrial system, scientific activities, and social institutions also characterized information processing and communication. The printing press, dictionaries, encyclopedias also reinforced the pattern of uniformity and specialization.

The success of the information-processing patterns of the industrial and scientific era, however, created new problems and challenges. The complexity of the numerical calculations required for further progress could no longer be handled with the available technology. As commercial and government administrative procedures became more data-intensive, the need to automate simple information-processing procedures became more critical. It was in this kind of environment that the motivation to invent computers arose. Once the technology of computers was available, information scientists started working on new information-processing techniques rather than just automating the precomputer techniques of the past. These new techniques addressed the problems of complexity and overspecialization. The general systems approach — or cybernetics — grew out of the need to control systems that had grown too complex to be handled by the information-processing techniques of the scientific era.

The scientific method succeeds by breaking down a system into its basic components and then narrowing the focus to one of those components so that a critical mass of research can produce a breakthrough or new paradigm. That paradigm is then articulated and applied in as many domains as possible (Kuhn 1972). Science analyzes systems into their basic components and deals with them one at a time in a linear fashion. This reductionist methodology creates a bias which sometimes makes it difficult for scientifically trained personnel to deal with a complex system such as the ecosphere, where the interactions of all of the components are so critical to understanding the overall operation of the whole.

The techniques of cybernetics allow an integrated systemic approach in which the impacts of each of the subsystems on each other are taken into account. The techniques of computers and cybernetics would never have occurred unless the complexity fostered by science had emerged. This parallels the rise of science as a way to deal with the complexity of the information collected with written and numerical notation. Whenever an information-processing system leads to information overload, it can only be handled by a new level of information processing.

Accompanying each breakthrough in information processing, there has been a major shift in education, as well as in the patterns of social organization and work. The following chapter examines the emergence of these new patterns and shows how the evolution of education, work, and social class parallels the evolution of language and information processing.

4

WHY THERE ARE SCHOOLS AND A MIDDLE CLASS: The History of Education and Social Class

Each step in the evolution of language affects the way a society sees the world and its economy. With the introduction of each new medium of communication or information processing, the education system, social class structures, and the organization of work have been transformed. I expect this pattern to repeat itself in our society as a result of the introduction of microcomputer systems in the late seventies into the workplace and the school system. To provide a perspective on the dimension and nature of the changes that microcomputers will bring, this chapter examines the role that communications and technology have played historically in the origin and evolution of education systems, social class structures, and work patterns.

These three aspects of social organization — education, social class, and work — are not independent of each other. The purpose of an education system is to train individuals to assume a productive role in society. The educational infrastructure of a society also influences the economic development of the society as a whole, hence the link between work and education. The tie-in with social class structure occurs because social class is a function of the way that work is organized in society and strongly correlates with the educational level of individuals and their families.

The institution of the school or formal education is such an intrinsic part of our culture that we take its existence for granted. We do not realize that schools, as opposed to education, are a relatively recent development. Schools were first organized only 5,000 years ago in Sumer, shortly after the invention of writing. The objective of these first schools was to train young people to become scribes. Education, as opposed to schooling, did not start with the advent of writing but can be traced to preliterate society. Preliterate education, however, is as different from formal schooling as oral communication is from written communication. Education and schooling are not the same, although the overlap during the industrial age was almost total. In the information age, education will once again be quite distinct from schooling. It will, however, be closely tied to work, as was the case in oral society, to the ongoing use of information technology, and to the process of lifelong learning.

In order to understand how social class structures might affect the organization of education in the information age, it is essential to examine how class structures have affected education in the past and, in turn, how education affected class structures. Class structure is not determined by how much a group earns but by how they earn a living. I will pay close attention to the middle class because of its close association with literacy and formal education. I want to explode the myth that the middle class is boring and show that the use of the term as a pejorative is unfounded. The middle class has historically been the source of innovation and revolution in society; in fact, it emerged as a result of literacy.

The appearance of social class is also relatively recent. Before the Neolithic revolution some 10,000 years ago, hunting and gathering societies were organized clans and tribes. A clan was an extended family, a tribe a small group of extended families working together to ensure their mutual survival. There was no such thing as class structure in these early societies; it is difficult to imagine members of the same family belonging to different social classes. Social class structures only began to develop with the introduction of agriculture, in which a division between the landed or upper class and the peasant or working class took place. The middle or managerial class followed shortly thereafter, with the introduction of writing.

Communication and Education

Educational practices changed radically with the introduction of written communication and continued to change each time a new mode of writing (for example, syllabic or alphabetic writing) was introduced. The invention of the printing press made mass education and self-education possible. In our own time, the electric media, such as motion pictures, radio, phonographs, audiotapes, compact disks (CDs), television, and video recorders which deliver audiovisual (A/V) information, have also had their own unique impacts, both direct and indirect, on education. Preliminary data indicate that computers will have a much more profound effect than any of the A/V media. Not only has technology affected education because of the way in which information can be delivered, but it has also affected what needs to be taught. As new information-processing techniques are developed, students must learn new skills. This has had an effect on the curriculum of schools as well as on how industry has responded to the training challenges of new technologies.

Understanding the relationship between communications and education will be essential for studying the impact that microcomputers are having and will continue to have on contemporary education in both the school system and in industry. Let us begin with the spoken word.

One of the chief features separating humankind from the rest of the animal world is our ability to communicate with each other through verbal language. Because of this ability, the knowledge and experiences that one human being acquires can be transmitted directly to others or stored as information. This process, independent of the medium of communication, is the basis of human education. Media of communication are not passive conduits of information. Rather, they are vortices of power, lively facilitators of social interactions and ideas. Communication media transform all aspects of human behavior, including an individual's cognitive processes and affective attitudes, as well as society's socioeconomic institutions of commerce, culture, and education. Understanding media and their effects are essential to understanding all human interactions and institutions, particularly those like education, which at its core depends on communication.

Communication permitted humans to share information and knowledge for their mutual survival and to plan activities based on this knowledge. Education arose naturally, as a means whereby parents or elders shared with younger members of their community information and knowledge that would increase their chances for survival. This form of education gave rise to distinct cultural communities among groups of individuals who were in day-to-day communication with each other and hence shared a common heritage and a similar set of experiences.

Communication between communities allowed for the growth and evolution of cultures and the amalgamation of smaller subcultures into larger cultural groups. Cultural evolution and development, education, and communication are inexorably intermeshed. One cannot understand one without studying the others. All communication is a form of education and all education is a form of communication. And both are manifestations of a society's culture.

Not only is communication the medium through which educational activities take place, but communication and the development of communication skills also serve as the focus or content of many educational activities. In other words, communication is both the medium and the content of educational activities. One of the primary purposes of education is to transmit the skills of effective communication. But language includes both a communications and an informatics component. Hence, another focus of education is the development of the information-processing skills associated with the use of language. As the complexity of human language grew and new modes of expression and organization evolved, the education system grew and evolved in turn. This chapter traces that growth from the humble beginnings of education within the oral tradition to contemporary schools in which all five modes of verbal language — speech, writing, mathematics, science, and computing — are studied and used.

The Oral Tradition

Education in preliterate societies basically takes two different forms. One is the vocational training of young people through their participation in the work activities of their parents, and the other is the practice of storytelling, whereby the values and wisdom of a culture are transmitted by elders to the whole society. Both forms of education are mediated by the spoken word, either through oral instructions in the former case or legends in the latter.

The basic form of preliterate education is mimesis or imitation, just as an infant learns to speak by imitating his or her parents. The oral instruction of vocational training was supplemented by demonstrations of how to do things correctly. The young people who were learning practical new skills did so by imitating the work patterns of their elders, whom they served as assistants or apprentices. Secret societies or fraternities formed to ensure the propagation of vital skills such as those of weavers, potters, toolmakers, warriors, hunters, fishermen, builders, and canoe makers.

Imitation also played an important role in the education provided by storytellers and priests. The tricks of their trade and the legends themselves were passed on through oral instruction and memorization. The exact words of each story, down to the accents, were memorized and transmitted through the ages in much the same way that nursery rhymes are preserved to this day. This is why the plot, rhyme, and meter were such an integral part of the stories of the oral tradition; they served as mnemonic devices aiding memorization. It also explains why the tales were always in the form of poetry and never prose. Prose was not a viable medium for transmission because its memorization would have been very difficult. The right-brain skills of pattern recognition and music were therefore an essential part of the education system and cultural development of oral society.

In addition to the wisdom and cultural values the storytellers imparted, they also provided listeners with certain tidbits of practical information. Woven into the plots of his tales, the storyteller would explain how to hunt for prey, or how to build a canoe, or how to conduct oneself in battle. Eric Havelock describes the bard as "at once a storyteller and also a tribal encyclopedia" who taught his lessons through the narratives of his tales (Havelock 1963).

CLASS STRUCTURES IN PRELITERATE SOCIETIES

Hunting and gathering societies are classless. Although they have leaders, decision making is largely achieved by tribal councils where each member of the tribe has input into the tribe's deliberations. It is only with the advent of agriculture and the emergence of landowners that a two-class society arose in which landowners functioned as the masters of peasants or serfs who became the underclass.

There are economic reasons for the advent of the two-class system. In a hunting and gathering society, it is difficult to preserve foodstuffs. Hence, the acquisition of wealth is impossible. Basically, food is preserved by sharing. When a family kills a large animal and there is more meat than they can consume, they share it with their neighbors or other members of their clan. This creates bonding between families so that when families that one shared one's meat with have a large kill, they will share their meat with you. Through this system of sharing, the overall survival of the cooperating tribe's people is enhanced. There is no survival advantage in being selfish in a hunting and gathering society.

With the advent of agriculture and the domestication of animals, the economics and politics of sharing changed. It was now possible to store food both through stockpiling grain and by maintaining a herd of domesticated animals. There was now a reason to acquire wealth and hoard or store food. From the point of view of the individual family, there was a reason for being selfish as it provided a hedge against starvation during bad times. There was also a reason to acquire land and seize ownership of it. The more land one could acquire, the more food one could produce and store, hence the greater one's wealth.

Preliterate agricultural society came to be divided into two classes — a ruling class of military, civil, and religious leaders and a governed class of serfs or peasants who performed the work that was required to grow the food and maintain the society. The structures of the first preliterate Neolithic cultures were quite different from those of a hunting and gathering society. Literacy and the associated formation of a middle class or third class did not emerge in agricultural societies until trading, commerce, and an urban form of civilization came into being 5,000 years later in Sumer.

From what we know of Sumerian culture, it was an absolutist society; power was shared between the aristocrats and the priests while the serfs were at the mercy of their rulers. One can only surmise that the aristocratic class obtained power through superior military talent and used priests to legitimize their power. The priests, through their knowledge of nature, were able to share power with the military elite. Thus, the two paths to power or exercising control in this society were either by being able to command brute-force military power, the route of the aristocrats, or by having an understanding of natural phenomena and by providing an explanation of them, the route of the priests.

The Sumerian culture prospered under this two-class system. It developed a surplus of foodstuffs which was used to support not only the aristocratic and priestly classes but also an artisan class and then a trading class. As a result of the activities of the commercial classes, a new system of information processing, and hence a tool for social organization, emerged in Sumer — writing. Those who developed a facility with literacy found that they were in a position to exercise influence in their society because of the uses to which writing could be put, not only in trade and commerce but also in the organization of the agricultural activities of the society. In Mesopotamia, irrigation canals were critical to the success of the crops. The construction and maintenance of the canals, therefore, required special coordination as well as a complicated system of accounting and land records to equitably distribute the life-giving water supply as well as the fruits of the labor of the farmers that used the water. As we have seen, it was the token system, employed to keep accounts of tributes paid in the form of agricultural commodities and units of labor, that evolved into the first form of writing.

Literacy and the Advent of Formal Education

There were no significant changes in the form of education based on apprenticeship and storytelling until the advent of writing in Sumer circa 3100 B.C. Shortly thereafter, there appeared the first formal education institutions which we call schools. "The Sumerian school was the direct outgrowth of the invention and development of the cuneiform system of writing, Sumer's most significant contribution to civilization" (Kramer 1959). Schools had to be organized to teach reading, writing, and arithmetic because unlike the other previous skill sets, these three activities cannot be learned from observation and imitation. Speech is the only form of language which is learned naturally by imitation. Formal instruction has to be organized for writing and mathematics. The secret codes of the letters and the numerals and their proper use have to be taught.

Infants begin to babble and imitate their parents' sounds automatically at a certain stage in their development; there is no instinct to scribble. Speech is acoustic and surrounds the young child. There is no escaping the influence of speech. Writing can only be observed if a child's visual senses are directed towards writing and what is seen is then explained. Being able to write requires a particular point of view that must be shared and explicated. The desire and ability to use speech as a form of communication is an instinctual trait built into all human beings independent of their culture. Information processing, the other aspect of language, is a learned trait acquired slowly and transmitted culturally from one generation to another through formal education. Without cultural institutions, information-processing skills are soon lost, as was the case in Europe during the Dark Ages shortly following the fall of Rome (with the exception of the monasteries). The capacity for speech, however, has never been lost, although different spoken languages have become extinct.

Judging from the exercise tablets prepared by students learning to write, along with descriptions of their training by their teachers found at the most ancient levels of archaeological sites, it is evident that schools were organized almost immediately after the appearance of writing. They were organized originally for vocational reasons, to train scribes for administrative duties. Reading and writing in education were at first used exclusively to teach reading and writing. As schools evolved, they were used to teach other subjects. Despite the practical motivation for the founding of these scribal schools, which appeared in every major urban center of Mesopotamia, scientific, scholarly, and literary activities began to develop in these institutions. Lists of natural phenomena such as trees, insects, rivers, and minerals; lists of geopolitical features such as cities and rulers; mathematical tables, lexicons, dictionaries relating different languages, theological tracts, and poetry were created at these schools and used as teaching tools (Kramer 1959). The "databases" inscribed on the clay tablets represented the first textbooks or software tools and were initially used to train students to learn the medium of writing; later, they became objects of study themselves.

The teachers in the scribal schools "were keen observers of nature and the immediate world about them. The long lists of plants, animals, metals, and stones which the professors compiled for pedagogic purposes imply a careful study of at least the more obvious characteristics of natural substances and living organisms" (Kramer 1959, 136). The information on the tablets was then used to teach other topics, including mathematics, science, foreign languages, poetry, theology, accounting, and administration.

We know a great deal about the earliest school system of the Sumerians from archaeological finds made in Iraq. Actual classrooms were unearthed in which clay tablets used to teach the children were found. The mode of organization of Sumerian schools set the pattern for schools right down to the present day. The shape of the rooms, rectangular; their size, accommodating thirty to forty students; and their structure, with rows of benches facing the teacher, bear an uncanny resemblance to our modern classrooms, which fundamentally teach the same subjects as were taught in Mesopotamia. Mesopotamian schools, like today's, were graded and hierarchical. There were primary schools in which basic literacy and numeracy skills were taught. There were also institutions of higher learning devoted to professional training. These institutes taught technical subjects such as astronomy, medicine, architecture, and engineering, or they specialized in the professions related to government service including theology (or temple service), law, commerce, teaching, and military science.

All societies which developed a formal education system possessed a writing system and all literate societies had schools. The ancient Incas are a possible exception because they had formal schools but did not have a true writing system. They did have a highly developed system of notation called *quipus*, which was a complex system of knotted colored strings that recorded both quantitative and qualitative information. The Babylonian, Egyptian, Hebrew, Vedic, Chinese, Mayan, and Aztec cultures are examples of the earliest cultures to have developed both a writing system and schools for training their young people.

The Egyptians were the second culture, after the Sumerians, to develop a writing system, and they were also the second culture to develop a formal school system. The Egyptians invented their writing system and organized formal schools sometime in the third millennium B.C. "The curriculum of elementary education was centered on writing" and writing was the "staple in all formal secondary education" (Power 1970).

In India, the Buddhists developed the institution of the monastery devoted to learning, which served as a model for monasteries in the Christian world, as well as for the Western university. In China, the education system served the government's need to train civil servants, who worked in the bureaucracy, the form of government that is itself a by-product of writing. The examination system was a key element in Chinese education.

An analysis of Hebrew education is particularly interesting because Hebrew culture is one of the oldest continuous literate traditions in the Western world, whose ancient history, including its oral tradition, is documented in the Bible and the Talmud (the commentaries on Scripture). The Hebrews were also one of the first cultures to make use of the phonetic alphabet. Their education system influenced Western culture in many ways because the three principal religions of the West — Judaism, Christianity, and

Islam — share a common origin and tradition beginning with the patriarch Abraham.

When they conquered and displaced the more sophisticated Canaanites, the Hebrews were a preliterate society with no formal education system. From the Canaanites, they adopted both the phonetic alphabet and an education system. The latter they transformed to accommodate their monotheistic religion. Originally, Hebrew education was family-based. Parents were required to teach their children "the testimonies, and the statutes, and the judgments, which the Lord, our God, have commanded" (Deut. 6:20). This passage was interpreted as a command to teach each child how to read so that he could read the words of the Lord and know His Commandments. The achievement of universal literacy by the Hebrews, an historic first, was motivated by religion but facilitated by the fact that the phonetic alphabet is the simplest possible writing system to learn.

In addition to home study, formal schools were organized to train scribes for government service and the study and administration of the Torah (law). The school system continued to grow and evolve particularly as the Hebrews encountered other cultures, usually as victims of military conquest. By the sixth century B.C., Hebrew education had evolved into "a universal and compulsory system of education" which operated until the destruction of the state by the Romans in A.D. 70 (Smith 1955).

Judging from its content, the Book of Proverbs, dating from the time of King Solomon, was likely used as curricular material for instructing the young. "Incline your ear, and hear the words of the wise, and apply your mind to my knowledge; for it will be pleasant if you keep them with you, if all of them are ready on your lips" (Prov. 22:17–19). This passage indicates that in Solomon's time, a mixture of oral and written learning was employed. The practice dates to ancient Egypt, where wise sayings were also an integral part of the formal school system. Oral sayings were used to impart wisdom about life to students, as well as for exercises that they could copy to practice their writing skills. Hindu education also applied a similar mix of oral and written learning, its students copying the Vedic texts and committing them to memory.

The Greek phonetic alphabet was the first to represent both consonants and vowels, making the learning of reading even simpler, and hence, as with Israel, a large reading public developed in ancient Greece and the Hellenistic states that succeeded it. The earliest writers in Greek culture were the pre-Socratic philosophers. The first of these, Thales and his followers, formed a school of thought, but it is not known if an actual learning institution was founded. Pythagoras formed a society to pursue philosophical studies and an institution of higher learning. Xenophanes also founded a school. These early institutions became models for the schools founded by Plato (the Academy) and Aristotle (the Lyceum). The Academy and the Lyceum set the pattern of scholarship for the next two millennia. Both emphasized the systematization of knowledge and the development of a methodology of learning based on rationality and deductive logic. This pattern, it has been claimed (McLuhan 1962; Logan 1986a), was influenced by the use of the phonetic alphabet and the spirit of analysis, classification, coding, and abstraction that it encouraged. "Plato encouraged his students to use logic, science, and rationality to find fresh solutions to problems instead of relying on the traditional remedies of the past. Together with a training in mathematics and logic, Plato encouraged his students to formulate the problems of human existence in scientific terms. The impact of the alphabet was to introduce both a new style of education and a new approach to problem solving" (Logan 1986a, 128).

Plato saw poets such as Homer as the enemies of his brave new world of rationality because they represented a powerful force for preserving the traditions of the past. They encouraged youth to imitate the patterns of the heroes of mythology, and hence, in Plato's view, prevented new patterns of behavior and thinking to emerge.

With Alexander the Great there was a change in the Greeks'

attitude towards education, likely due to the government's imperialism and militarism. The Greeks were the first society to impose their language and culture upon those they conquered. Education became an actual instrument of conquest. Alexandria, the world's first planned city, became the site of the Mouseion (from which the word "museum" comes), a gathering place for scholars housing a great collection of geological and biological specimens and the world's largest library. The local manufacture of papyrus enhanced the literary activities of the Mouseion. Manuscripts were collected from throughout the world and translated into Greek.

Defining the Middle Class

So far, the middle class has been referred to as the educated or literate class, that segment of society able to derive economic benefit because of the information-processing skills of its members, the class who often served as intermediaries between landlords and serfs. The term "middle class" as used here does not conform to the common usage of the term by most scholars. In fact, a precise definition has eluded social scientists. Most definitions involve strictly social, political, or economic criteria and as a result do not capture the essence of the middle class; they do not shed light on the reasons for the historical emergence of this group, which lie in their ability to exploit writing and other information technologies to advance their own and others' economic activities. Middle class is defined here as that sector of society which has readily been able to make use of literacy as a tool to advance its social and economic position in society. I will show through a reexamination of the history of Western education and information-processing techniques that, in fact, the middle class is the literate class and that the growth of the middle class and literacy are correlated. In addition, I will also examine the relationship of the middle class to the ruling class and the working class to determine how their position and role in society have changed over time. Finally, I will explain how traditional middle-class values — such as concern for education and stress on hard work and thrift — grew out of the middle class's intimate association with literacy.

As noted earlier, the middle class emerged along with literacy 5,000 years ago in Mesopotamia. Before the advent of the printing press, the middle class or those who could read were a very small minority of the population. The printing press created the conditions for mass literacy, and thus for the emergence of the middle class as a mass phenomenon and the major socioeconomic and political force of modern times.

It is increasingly clear that computers will have an impact on contemporary society as great as literacy had on previous generations. This impact could lead to the emergence of a new socioeconomic class, the computer class, which could replace the middle class as the vanguard class of the information age. In order to test this hypothesis, it is helpful to examine the relationship between the middle class and literacy.

One of the problems that has plagued students of social class has been their inability to arrive at a precise technical definition of the "middle class," despite its unambiguous meaning for the layperson. "Nobody has ever found a definition of the English middle class which is short, satisfactory, and watertight. To attempt to evolve one is to be lured into an interminable study of the social sciences" (Lewis and Maude 1949, 13). Commenting on the definition of the middle class as the "not-so-rich and not-so-poor individuals between the proletariat–labor group and the wealthy capitalist and social aristocracy" (Harkness 1931, 45), F.C. Palm remarked that "such a 'sandwich' definition indicates how difficult it is to characterize the middle classes. Anomalous, mutable, with tenuous fringes, the middle classes never have been and are not now a fixed entity, to be encompassed by a simple, rigid definition . . . Thus, the meaning of the middle classes is likely to remain with good cause in a state, so to speak, of suspended definition . . . a precise definition is virtually impossible" (Palm 1936, 3).

The term "middle class" was coined in the early years of the Industrial Revolution. "The first known mention of the term middle class is given by the Oxford English Dictionary as appearing in 1812" (King and Raynor 1981, 45). Webster's definition of the term, which reflects common usage and hence no particular theoretical bias, reads as follows: "middle class — the social class between the aristocracy or very wealthy and the lower working class." This definition reflects the etymology of the term as the class residing between the rich and the poor. The definition, however, provides no insight into the nature of the middle class, and therefore has been dismissed as inadequate by those social scientists who have tried to capture the essence of the middle class in a simple definition.

The middle class is often defined as the "bourgeoisie," which leaves us with the problem of defining what that is. Another equally vague definition is to identify the middle class as white-collar workers and the working class as blue-collar workers (King and Raynor 1981, 1–20) and (Roberts et al. 1977, 170). Still another definition associates the working class with productive labor (as in manufacturing) and the middle class with nonproductive labor (supervisory work) (Poulantzas 1978, 270). Closely related to these definitions is one which regards the middle class as nonmanual workers and the working or lower class as manual workers. "The middle class is often negatively defined as the residual category which remains when the manual working class has been subtracted from the population" (Gerrard et al. 1978, 1). Within the context of this definition, there are "three major approaches (Bain and Price 1972, 325–39) to defining manual and non manual work: the brain-brawn, the functional and the eclectic" (King and Raynor 1981, 1–20). According to the brain–brawn criteria, the middle class earn their living through their brains and the working class through their brawn. This somewhat pejorative definition is highly inaccurate because of the impossibility of separating, in many instances, the mental and physical components of a task. The functional approach defines middle-class activities in terms of four major job functions: administration; design, analysis and planning; supervisory/managerial; and commercial (King and Raynor 1981, 1–20). In the eclectic approach, the physical separation of the office and the factory are used to distinguish the middle class from the working class.

These three approaches to the manual/nonmanual criteria for distinguishing the middle class are not inconsistent with the definition of the middle class as the literate class, since literacy involves the use of brains rather than brawn, since each of the four functions defined as middle-class activities require literacy, and since literacy is used more often in the office than the factory. The question is to what extent the criteria are useful. As they are tied to an industrial form of the economy, their validity for analyses of the information age or the historical preindustrial era are not as valuable as the criteria of literacy.

Most social science definitions of the middle class occur within the context of a theory of social class operating within an economic system. The seminal thinkers in developing economic theories of social class were Karl Marx and Max Weber. Marx does not divide society into three classes — upper, middle, and lower. He confines his analysis to the capitalists who own the means of production and the proletariat who are exploited by them. The middle class appears as the bourgeoisie, defined as the capitalists and their agents. This conflicts with our hypothesis, since according to the Marxist view, the possession of a skill such as literacy does not determine class structure; the only thing that matters is the actual physical and political control of the means of production, that is, the factories. The weakness in the Marxist analysis of class is that its heavy emphasis on materialism and hardware does not allow for the consideration of information patterns or the social interactions of the information age. The control of the means of production in the age of information translates into possession of the skills of literacy or computer literacy or a combination of the two.

Weber identifies class in terms of economic interests and, in particular, the labor market. "According to our terminology, the factor that creates 'class' is unambiguously economic interest, and indeed, only those interests involved in the existence of the 'market'" (Weber 1946). However, he recognizes the role that skills such as literacy play in the way a class operates in the "market."

Of all the economic-driven definitions of the middle class, none is crasser than that of R. H. Gretton: "The Middle Class is that portion of the community to which money is the primary condition and the primary instrument of life" (Gretton 1919, 8). Other simplistic definitions classify the middle class in terms of certain brackets of earning power (Hulton Press Survey 1948, 19–20). "Except as a rough measure of what an income-earner has to spend . . . these brackets are inadequate to delimit the social classes in the nation" (Lewis and Maude 1949, 20). There are many instances when individuals who are clearly middle class earn less than well-paid laborers. Investigations with well-paid car assembly workers "revealed that even affluent skilled workers, whose relatively high income depended substantially on overtime and shift work, neither emulated middle class styles of life, nor mixed socially with non-manual neighbors" (King and Raynor 1981, 20–22). The middle class cannot, then, be defined strictly in terms of income or spending or other economic measures.

Another definition which incorporates the notion of education and technical skills reinforces the concept being developed here that the middle class is in fact the literate class. Giddens (Giddens 1981, 107) argues that advanced capitalistic societies are characterized by a threefold categorization of classes: upper, middle, and lower — based on three distinct sorts of market capacity, respectively: the ownership of property in the means of production, *the possession of educational or technical qualifications*, and the possession of manual labor power (King and Raynor 1981, 39–40).

Each of the definitions examined here has had some degree of validity, and thus some usefulness. Some come closer than others to my characterization of the middle class as the literate class, but all of them can be understood in terms of the notion that historically the middle class has been able to share power with the ruling class because of the information-processing skills they acquired through their literacy and educational background. In fact, a historical review will show that the origins of the middle class can be traced to the origins of literacy itself.

Literacy and the other skills to which literacy provided access enabled the middle class to dominate economic life, and hence to obtain either the best jobs or to start their own business or enterprise. This is the reason that the middle class has always enjoyed a better than average standard of living. Many of the members of the moneyed or upper class were actually the children (or more distant descendants) of middle-class entrepreneurs who became extremely wealthy. History is full of examples of aristocrats with middle-class roots, including the Medici, the Rothschilds, and the Rockefellers. In short, literacy has always translated into economic power, an observation that was noted by Claude Levi-Strauss:

> And when we consider the first use to which writing was put, it would seem quite clear that it was first and foremost connected with power: it was used for inventories, catalogues, censuses and instructions; in all instances, whether the aim was to keep a check on material possessions or human beings, it was the evidence of the power exercised by some men over other men and over worldly possessions. (Levi-Strauss 1969)

Acquiring literacy and the other forms of specialized knowledge based upon it has been the economic strategy of the middle class throughout history. "The middle class American shrewdly committed himself to the proposition that educated knowledge was the beginning of power, and power was the source of spiritual and material riches" (Bledstein 1976, 30).

Of all the definitions of the middle class, the one that comes closest to the ideas being developed here is that of Janet Coleman describing the emergence of the middle class in her book *Medieval Readers and Writers*: "There is an argument to be made in favour of defining a middle class in terms of literacy — which determines social rank — as well as with some contact with at least a market town, rather than defining them simply as those urban dwellers directly devoted to and in control of urban commerce and industry" (Coleman 1981, 63).

The Link Between Literacy and the Middle Class

If the hypothesis is correct that literacy was the tool which enabled the middle class to arise from the working class or peasantry, to prosper economically, and to evolve a unique lifestyle with an emphasis on education, thrift, and morality, then we should be able to find empirical evidence for a correlation between the existence of a middle class and literacy, as well as a link between literacy and the evolution of middle-class values. In fact, historical evidence reveals that the middle class emerged with literacy at the dawn of Western civilization and that there exists a one-to-one correspondence between increases in middle-class population and the level of literacy. The advent of the printing press, for example, marked a dramatic increase in both literacy and the influence of the middle class, two events which are not unconnected, as this chapter will show.

The middle class can be said to have first emerged with literate civilization as the group of self-made individuals with a rank lying between the nobility and the proletariat. They are also those members of society concerned with commerce, crafts, administration, and the priesthood. They served as a bridge between the upper class and the working classes, allowing the aristocracy to control the proletariat without direct contact. They organized and operated the irrigation systems that made large-scale agriculture possible. They collected tributes from the farmers so that they could support the irrigation workers as well as look after their own needs. They entered into trade so that the surpluses generated by the agricultural system could be stored for lean times or traded for other commodities of value. The ranks of the middle class were also filled by the civil servants who operated the administrative offices of the governments of the city-states and the empires into which the city-states evolved.

The existence of a commercial middle class in the rich Mesopotamian valley is amply shown by the law code of the Babylonian king Hammurabi (2067–2025 B.C.), which has numerous provisions dealing with merchants and the carrying on of commerce. Typical of such legislation is the following: "If the merchant has given to the agent corn, wool, or any sort of goods to traffic with, the agent shall write down the price and hand it over to the merchant; the agent shall take a sealed memorandum of the prices he shall give to the merchant" (Palm 1936, 8). What is interesting about this is the reference to writing. An integral aspect of the activities of the middle class or bourgeoisie from their very earliest history is an association with written records.

Direct evidence for the relationship of the middle class and literacy is provided by ancient records dating to 2000 B.C. in which some 500 individuals listed themselves as scribes, and provided the occupations of their fathers, as well, presumably for further identification. Their occupations as governors, city officials, ambassadors, temple administrators, military officers, sea captains, high tax officials, priests, managers, supervisors, foremen, archivists, and accountants (Kramer 1959, 3) clearly show the relationship between, or overlap of, the middle class and the literate class.

Ancient Egypt, the second civilization to develop a writing system, also had an active bourgeoisie. Egyptian artisans, merchants, and priests, most of whom were literate, accumulated wealth and pursued a middle-class lifestyle. Two other literate cultures arose in the territory between Egypt and Mesopotamia, each with an active middle class. One of these states was Phoenicia, "a nation of traders, [whose] citizens demonstrated astonishing initiative and persistent progress in domestic manufacturing, ship construction, foreign colonization, and general commerce" (Palm 1936, 10). The other was Israel, which was also a trading nation, particularly under King Solomon. Its most important contribution to Western civilization was not commercial but rather lay in its ethical writings and development of monotheism. The teachings of the Hebrew Scriptures have served as the paradigm for and the moral basis of the Western middle-class lifestyle and value system. For the larger part of its history, the Hebrew nation has adhered to a middle-class pattern of existence despite long periods in which it has endured poverty and other hardships. The Protestant work ethic of the North European middle class was strongly influenced by the teachings of the Old Testament.

Ancient Greece, the fount of Western civilization represents a culture where literacy and the middle-class way of life thrived. "A merchant class grew up, active, keen, and enlightened. It formed guilds . . . which stoutly defended their interests. It demanded written laws, and brought democracy into the constitutions . . . It even placed its stamp on the new intellectual life of the new Greece" (Glotz 1926). "Athens, during the Periclean democracy which followed, became a bourgeois state. With birth no longer a qualification for political preferment, the business classes played a foremost role in the selection of government officials. Many of the prominent leaders were of bourgeois origin . . . In cultural as well as in the economical life the Greek bourgeoisie played their part. Believers in education and individualism, they glorified Greek culture in all its forms, producing as a result many prominent members of the Greek intelligentsia from the middle classes" (Palm 1936, 12–13).

The role of the middle class and their lifestyle within the context of Periclean Athens was a pattern that repeated itself a number of times in the ancient world, including the Hellenistic empires, Rome, Byzantium, and the Islamic empires. Wherever urban centers developed within a national or imperial state, there thrived a literate middle-class society engaged in commerce, administration, and the arts, pursuing more or less the same goals and values as today's middle class.

Medieval Education and Class Structure

During the Roman period, the motivation for education and literacy was largely to pursue practical goals such as civil administration, commerce, or law. Literature and scholarship for their own sake were not as valued as they had been with the Greeks. Consequently, when the Roman empire collapsed together with trade within Europe, so too did the Roman administrative and legal infrastructure, along with the desire and motivation to be literate. Formal education in Europe fell into almost total decline, except for the activities of the Church and monasteries such as those of the Benedictines where sacred writings and some classical texts were preserved. Thus, when a nobleman or land administrator needed scribal services, he turned to the local clergy for assistance.

Although the monks were able to maintain the Western tradition of literacy, society in general reverted to the two-class system of a ruling class and a working class of serfs. The ruling class consisted of the civil authorities, the lords of the manor and other nobles (or members of the first estate), and the clergy (or members of the second estate). The peasantry or serfs formed the third estate of this society. The middle class in early medieval Europe disappeared. The Roman forms of organization couched in literacy and formal bureaucracies were replaced by the feudal system, which was based on personal loyalties in the form of fealty and oral oaths. These forms of social organization paralleled the tribal and oral traditions of the Germanic tribes that took political control of

Europe. The fundamental unit of economic activity was the manor, which, for the most part, was self-sufficient and did not engage in commerce or trade. There was no need for a middle class, and the only members of society that were literate were monks, isolated from everyday life.

Very little new literature was authored immediately after the Fall of Rome until the latter part of the Middle Ages. Beginning in the eleventh century, literacy entered the mainstream once again because of an increase in trading. Learning for the lay public was offered at cathedral schools, which were usually located along the principal trading routes. Starting in the fourteenth century, these schools evolved into the medieval universities which provided training in the arts, theology, law, and medicine.

Learning in the medieval university was determined and constrained by the need to create the texts that the students needed to study. The chief teaching methods were lectures, repetitions, and disputations. The ordinary lecture was usually nothing more than a dictation. The students copied the lecturer's dictation, which they then studied later. In a sense, the university lecture functioned as a means to reproduce texts deemed worthy of study. In some instances, the teachers were allowed to comment on or interpret their texts, but in many cases they were permitted only to perform straight dictation of the text. Repetitions in which the text was read out again were given so that students could check their copied texts for inaccuracies, although in some cases the text was also discussed. It was only at disputations that ideas were actually discussed and debated. The oral transmission of written texts was the only practical way of creating textbooks that the students of the day could afford. During classical times, slave labor was used to create books for those privileged enough for higher education, so the reproduction of books was not as great a concern as it was in medieval Europe.

Although the focus of study at medieval universities was on classical texts, a great deal of the learning occurred in the oral mode.

The content of the education system were written texts but the medium was the spoken word. The actual curriculum of the medieval university consisted of the seven liberal arts grouped in the trivium of grammar (literature), rhetoric, and logic, and the quadrivium of arithmetic, geometry, astronomy, and music.

Originally, education was open only to young men who were to enter the clergy. Later it was opened to include not only those who were interested in becoming clergymen, but also clerks who would become high officials under kings, including lawyers, judges, diplomats, scholars, or teachers. The numbers that were formally educated were extremely limited. The education experienced by the vast majority of young men was apprenticeship training in a trade, usually the same one as that of their fathers. The two-class social structure of the early Middle Ages began to change not only as a result of trade, but also because of technical breakthroughs that had a dramatic impact on agriculture and hence on the economy. The new technologies included the heavy plow, innovative techniques of irrigation and drainage, the harnessing of water and wind power, the stirrup, the cart, and harnesses for workhorses.

F.C. Palm attributes the reemergence of the middle class in the late Middle Ages to five factors:

1. A dramatic increase in agricultural production, which enabled a number of serfs to pay for their land from the proceeds of selling their surplus and thereby enter the ranks of the rural middle class.
2. The traders and craftsmen who served the needs of these new farmers organized their activities in market and fair towns and became burghers.
3. The Crusades promoted the interests of the middle class by opening new trade routes with the East.
4. The depletion of the ranks of the nobility due to wars, the Crusades, plagues, epidemics, and famines led to the distribution of their land to independent peasants.

5. The monarchs of the powerful kingdoms that arose during this period promoted the interests of and formed an alliance with the middle class to reduce the powers of the feudal lords and to promote the commercial and industrial interests of their kingdoms. The middle class were the vanguard of the new economic order of light industry and international trade (Palm 1936, 21, 25).

These five factors led to the emergence in the thirteenth and fourteenth centuries of "a well established, prosperous, independent middle class" (Palm 1936, 21, 25). "To complete the middle classes, there arose from the ranks of the simple artisans, organized mainly in the town craft guilds, the class of capitalist merchant-manufacturers" (Lewis and Maude 1949, 29–33).

These factors changed the nature not only of the middle class but of all three classes. There emerged in the upper classes two bases of wealth, power, and influence, namely, the land, as before, and now, with the coming of the Industrial Revolution, ownership of the means of production. The source of the new aristocracy of capital and industrialization was twofold: the old money of the landed aristocracy which was invested in new industrial schemes; and merchant manufacturers whose middle-class business operations grew into large-scale industrial enterprises creating wealth and influence almost overnight. A third source of the new industrial aristocracy came courtesy of the middle class; the inventors, scientists, and engineers and their technological innovations became the source of great wealth through the amplification of mass production. The ranks of the middle class increased with the addition of factory managers who served as liaisons between factory workers and owners. With industrialization, the working class or proletariat consisted of two distinct groups, the factory workers and the peasants.

Writing played an important role in the activities of the new middle class. It was used by merchants to keep records and organize commercial activities, and by municipal authorities to organize civic affairs. Writing was also used to express, through literature, the secular and moral concerns of the middle class. It is during this period that there was a major shift in the reading public from the first and second estates (the nobility and the clergy) to the merchant class. William Langland, author of *Piers Plowman*, was not supported by "a court patron but with the patronage of a devout 'middle class'" (Coleman 1981, 22).

There is a correlation between "the growth in lay literacy and social mobility as it was expressed in fourteenth century literature, a literature that did not merely passively reflect its time and context but was written as an encouragement to critique and change" (Coleman 1981, 17). Literacy not only became the tool which helped the middle class create a new economic niche for themselves, it also became the medium through which they rallied their causes and won political concessions from the first and second estates. "The literature of complaint, then, while seeming to express the grievances of the third estate, appears largely to be a literature by and for that most fluctuating estate, the wide-ranging 'middle class,' newly literate and newly vocal" (Coleman 1981, 64).

Renaissance Education, the Printing Press, and the Commercial Revolution

During the Renaissance, with its emphasis on humanism and individualism, many schools were founded to educate the young and make them literate. One of the factors that contributed to the growth of education and literacy was the greater availability and lower cost of learning materials in the form of books. This was made possible by two technological inventions — paper and the printing press. The arrival of paper in Europe from China had a major impact in reducing the costs of books. It also provided the motivation for the invention of the printing press because a supply of inexpensive paper to print on made the printing press a cost-effective technology.

The greater availability of books made it easier to operate a school and consequently more schools were established. Moreover, as more books found their way into society, there was greater opportunity for individuals to access reading material and hence additional motivation to become literate or to ensure that one's children became literate.

The printing press changed education by promoting the use of vernacular tongues rather than Greek and Latin, which had been the exclusive language of scholarship and learning. Indeed, until the printing press created a vernacular literature, children were taught to read Latin and not their own spoken language. Printers were eager to publish books in vernacular tongues because they needed to expand their markets in order to realize a return on the significant investment they made in purchasing their presses, type fonts, and other equipment. The absolute number of books and the percentage of books printed in vernacular tongues as opposed to Latin and Greek increased markedly. A vernacular literature also increased the number of individuals who could learn how to read and write since they did not have to learn Latin first.

While it was in the late Middle Ages that the middle class first established itself as a permanent fixture of Western society, it was only during the Renaissance and the Industrial Revolution that followed that this class grew into a popular or mass movement, with the help of the printing press.

One of the features of the commercial revolution during the Italian Renaissance was the increased use in business affairs of literacy and organizational techniques that grew out of alphabetic literacy, such as the double entry system of bookkeeping, the bill of exchange, and other banking instruments. A measure of the influence of this period is the current use of business terms derived from the Italian words *disconto*, *netto*, *deposito*, *foglio*, and *bilancio* (Palm 1936, 32–33).

There was a practical side to education, as well. The bourgeoisie considered education "a means of fostering the moneymaking powers of their sons, for as bankers, traders, manufacturers, and landowners, they could readily see the practical value of education; consequently, children were hastily schooled in the rudiments of business and put directly to work at the occupations of their fathers" (Palm 1936, 38).

The commercial revolution fomented by the emerging middle class also had an enormous impact on cultural patterns, including the revival of learning and the championing of the individual. Once firmly rooted economically, the middle class sought outlets for the expression of their wealth. Not only did they become patrons of the arts like the nobility, they also participated directly in the production of works of art and literature. "Italian middle class people tried to attain social distinction by personal achievement in art and literature as well as good manners. Most of the Renaissance artists and scholars were of the business classes and proletariat" (Palm 1936, 39).

With the reemergence of the middle class in the thirteenth and fourteenth centuries, there were now two literate sectors in society, the merchants and the clergy. The members of the original first estate remained largely illiterate, but with the advent of new forms of social, economic, and political organization, a change occurred in the makeup of the ruling class. This development resulted partly from the upward mobility of the literate middle class who entered the ruling classes by virtue of their commercial successes, like the Medici family. Those upper-class families whose ancestors arose from the middle class often retained their tradition of and interest in literacy.

This phenomenon may give rise to some confusion in terms of defining the middle class as the literate class because of the existence in it of literate members of the upper class. There is nothing exclusive about these categories, however. If we define a member of the ruling class as one who is able to exercise control in society by virtue of wealth or power and a member of the middle class as one who is able to exercise control in society by virtue of facility with literacy and all the skills associated with it, the conflict disappears. These two categories are not mutually exclusive. It is possible for a member of the upper class to have all the characteristics of a member of the middle class and to adhere to that value system. Since this definition does not refer to earning power or level of wealth, an upper-class person can also be middle class, just as a struggling artist or writer can be extremely poor and middle class at the same time. Some of the confusion in class structure analyses of society occurs because classes have been defined in terms of their financial situation rather than their function within society.

Since I do not define a class in terms of earning power, I must also provide a definition of the working class. A member of the working class is a member of neither the middle nor upper class, and hence is unable, due to lack of wealth, power, or education, to exercise any level of control in society, even over one's own existence, except by virtue of selling one's labor. This is the reason that a moderately well-to-do member of the working class cannot really be considered a member of the middle class. My division of classes, then, is in terms of the power base of the individual in society. For the upper class, this is through wealth or power, and for the middle class through literacy and education. Unfortunately for the working class, their power base is limited and confined to collective action through the trade union movement or through political revolution.

Beginning with the humanist movement at the beginning of the Italian Renaissance and continuing through the commercial revolution, the Enlightenment, the rise of democracy, the American Revolution, and the French Revolution, the middle class fought a steady battle with the aristocracy to win a greater share of the political power and wealth of their society. They did this by mobilizing politically through the use of the written word and the printing press, proving Voltaire's adage that the "pen is mightier than the sword." In the commercial world, they were able to improve their well-being materially by implementing the latest techniques of business organization based on literacy. In the industrial world, they prospered because their superior level of education allowed them to take advantage of the latest developments in science and technology.

As a force within society, the middle class began to play an increasingly important role in the affairs of European society in the late Middle Ages. As the technological base of economic activities increased, so too did the influence and importance of the middle class. Also, the ranks of the upper class changed dramatically, for two reasons: the continuing social migration of the prosperous middle class into the ranks of the upper class; and a migration in the opposite direction as members of the old aristocracy joined in the literacy revolution and adopted some of the values of the middle class through acculturation.

Interactions between the middle class and the upper class have remained much the same to this day. The working class, however, have been able, beginning in the nineteenth century, to change their position in society through the trade union movement. This, along with the increasing affluence of Western industrial civilization, has permitted them to earn a decent standard of living in the past century, at least in Western Europe and North America.

THE RISE OF NATIONALISM

The availability of books made possible by paper, the printing press, portability, and a vernacular literature made possible another educational phenomenon, the self-learned man, an individual who educated himself by reading a great variety of books. One such was Tartaglia, who translated Euclid's *Elements* into Italian. Those who learned informally through their readings — practical men involved in the commerce of their society — began to rival the universities in terms of learning. A society of men interested in developments and learning began to develop in the late fifteenth century around the print shops of the major cities of Europe. As the publishers of new ideas, the printers were in a better position to be current with the latest developments in science, technology, commerce, and scholarship than the professors at the universities who were bogged down in the traditions of the Middle Ages. Printing presses were first set up in the commercial, not the university, towns of Europe with the exception of Wittenberg. Knowing how to read soon became an essential feature of commercial life. By the sixteenth century, the acquisition of writing had become "absolutely necessary in virtually every aspect of middle-class life" (Bowen 1972, 402).

One of the consequences of the creation of a vernacular literature made possible by the printing press was the rise of nationalism. Nationalism, in turn, gave rise to the desire for national education systems. The printing press made this goal a reality by allowing uniform national curricula based on mass-produced textbooks to come into existence. "Henry VIII could require all schools to use a single textbook in grammar and no other" (Cole 1950). The use of the same textbook at the national level also led to a uniformity of language and the notion of "correct" grammatical uses of vernacular tongues in both their written and spoken forms.

PROTESTANTISM, EDUCATION, AND THE MIDDLE CLASS

The printing press had an indirect impact on education through the Protestant Reformation of the early sixteenth century, a movement which was closely tied to this technology. The new style of commerce, which may be characterized as capitalism, came into conflict with the teachings of the Roman Catholic Church, which was opposed to usury and excessive profits. During the Middle Ages, the church had supported the guild system of a just price, which regulated and protected the trade of the middle classes. Now this class was finding itself restricted by the Church. The result of this conflict was to fuel, to some extent, the Protestant

Reformation (Weber 1948). Naturally, other factors contributed to this revolt, but these were also tied up with the social and political aspirations of the literate middle class. Their desire for political freedom and the expression of their individualism contributed to the social ferment that led to the Reformation.

Literacy also played an important role in the evolution of the Protestant Reformation. Protestants were encouraged to read the Bible on their own. It was in the Protestant territories that printers began to publish vernacular forms of the Bible. In fact, even before the Reformation began, the first use made of the Gutenberg press was to print the Gutenberg Bible. The printers also played a key role in spreading the reforming ideas of Luther and Calvin through the many broadsheets and pamphlets they published. Luther's famous 95 Theses which he had nailed to the church door in Wittenberg "were said to be known throughout Germany in a fortnight and through Europe in a month" (Aston 1968), thanks to the services of the printing press.

Empirical evidence for the thesis that the emergence of the middle class, mass literacy, the printing press, capitalism, and Protestantism are interrelated and mutually reinforcing is found in the historical correlation of these activities in centers such as Holland and Geneva. "With the victory of the Calvinists in Holland the Golden Age of Dutch printing had dawned" (Eisenstein 1979, 409). Holland was not only a great center for Calvinism and printing, it was also a haven for trade and commerce where the middle class prospered and democracy grew strong.

UNIVERSAL EDUCATION, THE INDUSTRIAL REVOLUTION, AND THE RISE OF THE MIDDLE CLASS

The evolution of education systems into a universal and compulsory system was due to the Industrial Revolution and the factory system of mass production. The factory not only permitted the mass production of goods, it also created wide-ranging social changes. For example, it triggered a mass migration of families from farms to towns and cities by people seeking work in the factories. At first, at the beginning of the eighteenth century, all the members of a family — children and adults — worked in the factory. The working conditions were deplorable until, after a public outcry, child labor laws were passed preventing children from working in factories. This created another social problem of equal magnitude: there was no longer a role or useful activity for working-class children in urban society, unlike the situation on the farm, where they could perform chores, help their parents, and be properly supervised. Elementary schools offering universal education to the young were organized in England in 1870 to address this social need.

Alvin Toffler has suggested that mass education evolved as it did because many of the children of the publicly supported school system were destined for factory work once they became adults. Schools were organized to train them for that activity. Toffler writes: "Built on the factory model, mass education taught basic reading, writing, and arithmetic, a bit of history and other subjects. This was the overt curriculum. But beneath it lay an invisible or covert curriculum that was far more basic. It consisted — and still does in most industrial nations — of three courses; one in punctuality, one in obedience, and one in rote repetitive work" (Toffler 1980, 45). These lessons were intended to make sure that workers showed up on time, took orders from their bosses without questioning, and stayed on the job "performing brutally repetitious operations" all day long.

Within the new industrial order, work was no longer individualized as it had been during the Middle Ages when production was based on craftsmanship. The education system that developed during the Industrial Revolution reflected the mass-production assembly-line style of production. The factory served as a model for education, which was delivered to a mass audience in assembly-line fashion. This was not true for universities and the private schools which had been founded earlier and were attended by the privileged members of society, but it was certainly true for the public school system which was organized to serve working-class populations. The modern industrial state took over education to ensure that its economy would benefit from a uniform, productive working force.

The uniformity that the printing press and industrialization encouraged also influenced how children's education was planned and their performance evaluated. Unless a child was able to learn at exactly the same rate as his peers and advance through the curriculum lockstep, he fell behind and was demoted or demoralized, often both. As we will see, one of the refreshing aspects of the microcomputer revolution is that it reindividualizes education and retrieves many of the patterns of education of the oral tradition and the medieval era. It also reindividualizes work and provides scope for personal input and achievement.

A major step in the growth of the middle class and their influence was the Industrial Revolution. The invention of the steam engine in 1769 by James Watt allowed the full mechanization of manufacturing and greatly improved transportation through the steamboat and the steam locomotive. The application of science and technology to agriculture, medicine, and sanitation also contributed dramatically to an increase in population and an improved standard of living. More people were capable of supporting the middle-class lifestyle. "It was the Industrial Revolution which gave the middle class a new self-consciousness and for the first time the sharing of a common identity which seems to have lasted to the mid-nineteenth century" (King and Raynor 1981, 48).

The Effects of Electric Media on Education

Beginning with the discovery of electricity in the mid-nineteenth century, a host of inventions have influenced education, including the light bulb, the telegraph, the telephone, the phonograph player, moving pictures, the microphone and loudspeaker, radio, television, slide projectors, overhead projectors, audiotape and videotape recorders, hand-held calculators, fiber optics, compact disks, and optical disks. The effect of these devices on education is quite different from that of computers.

To properly assess the effects of the A/V media, a distinction must be made between these media and the interactive media of written language and computers. Users of A/V media are passive consumers of the information, sounds, and images that these media deliver. The users of both written language and computers, whether they are receiving or transmitting information, must interact with that information through a process of coding (if they are writing or inputting data) or decoding (if they are reading or accessing data). These media therefore stimulate cognitive development in a way that A/V media cannot.

The educational value of A/V media is restricted to their being a source of information, whereas written language and the computer are a source of information as well as a medium for organizing and processing information. Interacting with the written word on a computer contributes to the cognitive development of the user in a way that watching television or some other A/V medium cannot. Even reading, because of the decoding it requires, involves an organization of the material presented. The reader is constantly deciphering, understanding, interpreting, editing, rereading, or studying the material that is presented.

Nevertheless, the use of A/V media has a number of important educational implications. Their use is not without some cognitive impact. The cognitive skills that are stimulated by A/V media are similar to those activated by the reception of oral communication. Perhaps the most significant impact of A/V media on education is not the way in which they stimulate cognitive development but the manner in which they compete with, interact with, or complement the more interactive media or languages of writing, mathematics, science, and computing.

The general impact A/V media have had on society has already been discussed in Chapter 2 and their effects have been contrasted with those of the written and printed word. The observations made there also apply to education and can be summarized as follows: The written and printed word encourage patterns of information processing and organization which are linear, sequential, analytic, specialized, and logical. Electric and electronic media, on the other hand, encourage information-processing patterns and organization which are holistic or global, integrated, synthetic, generalist, and metaphorical.

Changes in the education system began to occur with the invention of the telegraph in 1844. The dimensions of the world suddenly shrank to those of McLuhan's global village. The study of other cultures and nations took on new meanings as news from distant events arrived by telegraph almost as they happened and were reported the next morning in the newspaper. An invention as ostensibly simple as the light bulb also had its effects. It made possible individual study late into the night, as well as night school for those who had to work during the day. The telephone, like the telegraph, increased interpersonal communication over long distances and brought the world closer together. Radio, movies, slide projectors, phonograph records, audiotapes, and television provided listening-and-viewing audiences with a rich panoply of sights and sounds gathered from across the world, reinforcing the notion of a global village. These media also provided important new channels for the delivery of information that did not depend on text. One implication for education was that many phenomena which were difficult to describe verbally could be experienced directly thanks to A/V media.

Television provides a fascinating example of the complex relationship a medium can have with education. At first, television was touted as a medium that would have a major positive impact on education. Now that we have experienced more than forty years of commercial television and every viewer has access to one or more educational channels, we are in a better position to evaluate television's influence on education. It has certainly had a major impact, but not always what was expected and not always positive.

Some of the promise of television has been realized. It is an excellent medium for the dissemination of information and for giving the viewer a sense of "being there." Educational television has provided, on balance, informative and entertaining fare, but there are limits to how effective it has been. For example, *Sesame Street* has helped correct some of the negative racial stereotypes which abound in North American society, but it has not achieved all of its literacy and numeracy goals for ghetto children.

My quarrel with television, if I have one, is not with the failure of the dream of a positive contribution; rather, it is with the extremely negative influence television has had on education, particularly the education of youngsters. By providing information and entertainment in such an easy and effortless manner, television has discouraged children from reading more. Television occupies such a large chunk of the leisure time of children, it robs them of the opportunity to spend more time reading, takes away their desire to read, and destroys their attention span. Too much effort is required to obtain, through reading, the same amount of information and excitement that can be obtained effortlessly from watching television. Television provides children with carefully digested and packaged nuggets of information in extremely short time intervals of approximately ninety seconds in length. Because television destroys children's attention span, the painstaking process of learning the three Rs has been made more difficult.

Luckily, however, the computer has reversed some of children's disinterest in learning and stands poised to undo some of the harm that television has incurred.

5

THE MICROCOMPUTER IS THE MEDIUM AND EDUCATION IS THE MESSAGE: Education, Work, and Computing

Discussion of the evolution of communications, information-processing technology, and language and the ways they have influenced the development of social institutions has provided background for understanding the enormous changes computing will have on contemporary social institutions. The next step is to examine the impact of computers on work and education.

The Computer, An Extension Of the Human Psyche

The character of data processing has been completely transformed because of the speed with which simple operations can be performed using computers. Each of the individual steps of information processing done by a computer can also be done by the human mind. What makes the computer so powerful is the speed (orders of magnitude faster than human thought) with which simple calculations and/or operations can be processed. This makes it possible for the computer to perform certain tasks that would otherwise be impossible through human thought because of time limitations. The computer perfectly illustrates that a quantitative change can become a qualitative change if it is large enough.

I have described the computer's capabilities as if "the computer performs tasks." This description tends to anthropomorphize the computer and may give the reader the impression that the computer possesses a form of intelligence. The use of the word "intelligence" is an unfortunate shortcoming of our limited vocabulary which was not designed to deal with machines capable of automated processes. To be more accurate, one should talk of the computer performing tasks *as directed by human intelligence*. As it is tedious to add the qualifier, "as directed by human intelligence," each time I mention the "performance" of the computer, the reader is asked to keep this qualification in mind.

This distinction is not a linguistic quibble but an important pedagogical and philosophical point. The computer is not a tool that thinks for us; it is one that extends human intelligence. Moreover, it is not the first tool to do so. Computational devices such as the abacus and the adding machine, a primitive forerunner of the computer, are examples of technologies that enhance human intelligence. Information-storage devices such as books, encyclopedias, and libraries are examples of other tools which also extend human intelligence. The computer incorporates the computing capability of calculators and the information-storage capability of print or written records.

By increasing the amount of information that the human mind can comfortably access and use, the computer creates new patterns of information usage and facilitates the process of discovery. Because of the computer's power in creating new knowledge and solving problems, a notion of artificial intelligence has emerged which can be misleading. The danger is, we may begin to allow computers to do our thinking for us and make decisions based on preprogrammed instructions. Many of us, no doubt, have encountered situations where employees of a large organization are forced into what seems to be a stupid decision because their computers provide them with no other alternative. To avoid such situations, it is essential to regard the computer as an extension of human intelligence directed by human volition. The designers of computer programs should make more allowances for human interaction whenever the computer forces its users to violate common sense.

Not only is the processing of information enhanced by computers, so too is the storage and retrieval of information. The computer is able to store extremely large quantities of data. This is also true of a book or a library filled with books. The essential difference, however, is that because of its processing capabilities the computer can access information rapidly. The computerized search for and retrieval of information represents a quantum leap over traditional paper methods. The ability to search for and link information allows new forms of information processing and organization to emerge such as the electronic spreadsheet, the relational database, and word processing. The quantitative changes in information storage and retrieval due to computers create qualitative changes and hence warrant our consideration of computing as a language different from the other forms of notated language — writing, mathematics, and science. The storage of information by computers also has the added convenience that information can be retained in the computer's memory and then accessed, as needed, in either hard copy (a paper form) or on a video screen.

Computers have also transformed the organization of information. While the forms of activities and functions encompassed in desktop publishing, word processing, database management, and spreadsheet analysis existed in purely paper forms, the existence of computers has transformed these four activities. Before computers, the activities encompassed by desktop publishing and word processing were confined to typewriters, traditional typesetting, and scissors and pasting or other mechanical layout procedures; database management relied on mechanical indexing and cataloging techniques; and spreadsheet analysis consisted solely of standard bookkeeping techniques enhanced by the use of adding machines and other electromechanical computational devices. Word processing, desktop publishing, database management, and spreadsheet analysis have become the principal forms of information organization for the work world in the computer age. They dominate office procedures and are carried out primarily with the help of microcomputer systems.

Word processing, desktop publishing, databases, spreadsheets, and general systems theory are specific techniques that the computerized organization of information encourages. Each of these techniques in turn encourages a style of organizing information which differs from manual systems and is perhaps best characterized as integrated. The integration occurs at many levels. First, there is the integration of qualitative and quantitative data, or put simply, the integration of numbers and words. The computer, as its name implies, began its existence as a high-level calculator, a number cruncher. Because the computer is actually a manipulator of abstract symbols, whether they are numerals or letters of the alphabet, it also became a manipulator of text very early in its existence. In fact, one of the first applications of computers during World War II was the deciphering of intercepted coded messages. Spreadsheets and databases are excellent examples of the way in which computers integrate numerical and textual data.

Not only are the processing, storage, retrieval, and organization of information transformed by computers, so too is the communication of that information. The computer has three basic forms of outputs by which it can disseminate information. These include the traditional print format through the use of a variety of printers (dot matrix, daisy wheel, or laser), a video format through the use of cathode ray tubes (CRTs) or light emitting diodes (LEDs), and as electronic digital signals (used primarily for computer-to-computer communication). Modems connected by telephone lines allow computers worldwide to swap information instantaneously.

THE NATURAL LANGUAGE FOR DATA INTEGRATION AND HYBRID SYSTEMS

The use of computers to process graphical and musical elements of human expression is also growing. Specialized software packages exist which allow computer users to create still graphical images for either print or for projection slides (35mm or overhead); to scan existing images and edit them to create new images; to create engineering and architectural drawings (CAD); to create animation; to edit video and motion picture film; and to synthesize, compose, and notate music.

Desktop-publishing software, which allows the user to create professional-looking page layouts, has transformed the nature of publishing and printing. Desktop-publishing software which integrates text and graphics has served as a model for multimedia software which integrates graphical, video, and audio material together with interactive mechanisms. Multimedia programs allow the user to interact actively with information presented in a variety of forms rather than making the user a passive consumer. Another application is a multimedia educational package in which a software package combines textual material, including interactive questions, with computer-simulated material or video and audio material presented by a videodisk or some other video medium (Hannifin and Hughes 1985, 236). The possibilities are endless. A new class of media known as hypertext and hypermedia has been created in which textual information or, alternatively, multimedia, are connected within a relational database so that there are multiple ways of accessing and linking information. A hypertext system is a database in which information has been organized into smaller units of information called nodes; these nodes can be easily joined to other information nodes within the database. Hypermedia is an extension of the hypertext concept in which the nodes of information are in multimedia formats rather than text. Hypertext and hypermedia, because of their distinctive organizational principle and methodology of user interaction, constitute a new and distinctive medium with the potential to radically change both our education system and the workplace. Its use on the Internet is growing, particularly on the World Wide Web.

The computer allows the integration of a number of information-processing tools because the output from one software package can be fed into another as input and then passed on to still another package for further processing. Computers provide this level of integration because of their ability to network different users and systems through the use of local area networks (LANs), wide area networks (WANs), and the Internet. Not only does networking enhance communication, it also facilitates information processing and organization by making data that reside in many different computers available to a single user. In principle, all computers could be connected through modems and the worldwide telephone network, creating a "global computer," an analog to McLuhan's notion of the global village. The Internet is rapidly making this possibility a reality.

The Paradigm Shift from Era I to Era II Computing

The evolution of computing is usually described as a continuous process. Dan Tapscott and Art Caston (Tapscott and Caston 1993), however, see a discontinuity in the way in which computing developed. They describe a paradigm shift in the use of information technology from the original exclusive use of mainframe computers to the current mixed use of networked mainframe, minicomputers and/or microcomputers typically configured as distributed client/server systems in which a host computer is serving information to computer workstations distributed throughout the workplace. Tapscott and Caston labeled the initial period of computing dominated by mainframes as Era I of the information age and the more recent period dominated by microcomputers as Era II.

Era I computing began just after World War II with the development of the mainframe computer, first with electric relays, then with vacuum tubes, and finally with transistors. Initially, only technically trained computer experts performed calculations for those requiring computing services. Eventually, end users were allowed to use the computer by submitting jobs on punched cards that were batched together with other jobs and were read into the computer with a card reader. The computer would process these jobs together and send the results to a centralized printing device. After the jobs were printed out, they were distributed to the users. The punched card and centralized printer system was eventually replaced with terminals that allowed the user direct communication with the computer. The next development in computing was the minicomputer, which greatly reduced both the scale and the cost of computing. A greater range of applications became affordable. Minicomputers could be dedicated to specialized tasks and therefore led to some decentralization of the computing function in large organizations.

The next breakthrough was the introduction of microcomputers, or personal computers, which created the revolution. They democratized computing totally because they lowered costs dramatically and have continued to do so since their introduction in the late '70s. They also provided a computing environment that could be dedicated to the needs of the individual user, hence the appellation the personal computer. The microcomputer gave individuals freedom from costly and hierarchical mainframe and minicomputer solutions which did not always serve the end user's needs.

The breakpoint between Era I and Era II information technology was created by two developments: first, the emergence of standalone microcomputers, and second, their integration with mainframes, minicomputers, or other micros to form distributed networks or client/server systems in which the host computer downloads software and information for processing on local microcomputers. The first step in the development of client/server systems was the creation of local area networks (LANs) of micros through the use of communications cards inserted into the microcomputers together with specialized communications software. The second step was the creation of a similar linking between micros and a mainframe or mini host computer. The purpose of LANs was to allow information to pass between users and thereby recapture some of the advantages of mainframes. The linking of a host computer with micros was motivated in part by the desire to reintegrate the community of computer users after the disintegration caused by the use of stand-alone micros.

The creation of client/server systems achieved these goals and provided microcomputer users with a richer computing environment in which resources such as software and databases could be shared by many users. The client/server environment is superior to the former host-based systems because individual users in the network are no longer tied to a centralized computing authority. They are free to pursue their own development to meet their individual needs and at the same time take advantage of shared resources. The Internet system, which will be described in greater detail in Chapter 8, is an example of a global client/server system in which users from around the world dial into a virtual host consisting of all the Internet nodes to share information and software files.

EIGHT CRITICAL TECHNOLOGY SHIFTS

Tapscott and Caston (Tapscott and Caston 1993) describe the transition from Era I of information technology to Era II of information technology in terms of the following eight critical technology shifts:

Shift 1: "From Traditional Semiconductors to Microprocessor-Based Systems." This shift allowed for the miniaturization of computers so that they could fit on the worker's or student's desktop. The microcomputer permitted a dramatic reduction in price that made their use in business extremely practical and their use in schools and for personal development affordable, hence contributing to the democratization of computing. It also contributed to the decentralization of computing power and thus the general leveling of industrial-age hierarchical structures.

Shift 2: "From Host-Based to Networked-Based Systems." This shift also contributed to the decentralization of computing power and the leveling of hierarchical structures. It provided a hardware environment for the integration of communications between computer users in the workplace and in the schools.

Shift 3: "From Vendor Proprietary Software to Open Software Standards." This shift enabled each organization to develop its own affordable computing environment without being held hostage to arbitrary pricing or the technical expertise of a third party.

Shift 4: "From Single to Multimedia: Data, Text, Voice, and Image." This development enriched the information environment that the computer user could access and integrated visual, aural, oral, and textual information.

Shift 5: "From Account Control to Computer Vendor–Customer Partnerships Based on Free Will." As with open systems, this shift allowed organizations greater freedom in developing their own systems.

Shift 6: "Software Development: From Craft to Factory." This shift greatly reduced the cost of software and permitted the development of standards and generic skills. It provided a basic set of information-processing tools consisting of word processing, database management, spreadsheets, presentation, and desktop publishing which could be universally accessed and used interchangeably between organizations. It has now provided learning tools for use in the classroom that allow children to discover in a systematic and relatively simple manner basic generic information-processing skills.

Shift 7: "From Alphanumeric to Graphical, Multiform User Interface." This shift has also contributed to the democratization of computing by allowing many more people to access computing. The graphical user interface makes software easier to use and faster to learn. It lowers the age at which children can begin to use computer applications. It integrates analytic data-handling techniques characteristic of left-brain processes with spatially oriented strategies of information processing characteristic of right-brain processes. This allows children and adults suffering from dyslexia greater access to computers and a viable strategy for overcoming their cognitive deficits.

Shift 8: "From Stand-Alone to Integrated Software Applications." This shift, together with network-based hardware, has permitted the integration of communications and information sharing among computer users.

The shift from Era I to Era II computing has profoundly affected the way computers are used in schools and the workplace. The difference in education is, perhaps, more dramatic because Era I computing had almost no impact on education. There were a few isolated applications of computer-assisted learning which merely automated learning and teaching strategies from the paper-and-pencil era. The only really significant use of computers was as the object of study in computer science courses delivered in high schools and postsecondary institutions. Era II computing, on the other hand, stands poised to create a revolution in education.

The shift from Era I to Era II information technology has been dramatic in the workplace, as well. It has resulted in completely new patterns of organizing work. Contemporary business is currently in the process of carrying out this reorganization, which frequently goes under the name of business process reengineering. The reorganization entails new patterns of work for the individual worker, but perhaps more importantly, promises the greater integration of business processes and communications both within and between organizations.

The eight shifts listed earlier are resulting in the completion of the shift of the pattern of work from that of the industrial age to that of the information age. But the way the use of computers has been organized in the workplace, ironically, owes more to the patterns and organizational principles of the industrial age than other segments of the business world; in the case of software production, the pattern has resembled the preindustrial era of crafted technology characteristic of the Middle Ages. In order to understand this paradox, let us return to the communication theories of Marshall McLuhan.

A MCLUHAN PERSPECTIVE ON ERA I AND ERA II COMPUTING

Chapter 2 examined how, beginning with the telegraph, the electric flow of information created new patterns of work (McLuhan 1962; 1964). By the time McLuhan first formulated his ideas in the early sixties, the technologies of the telegraph, telephone, radio, television, and photocopier had begun to break up the specialized, centralized, hierarchical structures of industrial age business into more decentralized, more integrated, and flatter organizational structures. McLuhan's analysis of the effects of electric information led him to predict that work patterns would also be characterized by decentralization, nonhierarchical, generalist, and integrated structures.

How is it that McLuhan was able to observe the organic, decentralized, and integrative patterns of Era II of the information age a full twenty years before the introduction of microcomputers, in the heart of Era I, when the use of mainframe computer technology was characterized by the opposite patterns? McLuhan understood that when a medium first appears, it uses the content of another medium exclusively for its content until its users have learned to exploit the new medium to develop new forms of expression. In addition to this phenomenon, the nature of the mainframe computer must be taken into account.

Like the factory, the mainframe computer represented a considerable capital investment and therefore its use had to be organized to increase the efficiency of the equipment. As with factory workers, shifts were organized to minimize equipment or machine downtime. Jobs run on the computer were organized in batch modes to maximize the machine's throughput. The user had to organize information processing to accommodate the needs of the computer and often the high-paid staff that operated and programmed it.

The first uses to which mainframe computers were put were to gather information and create reports identical to those that had been created before the advent of computing. Instead of creating new business procedures, computers were used to automate the old tried-and-true methods for conducting business. Moreover, they provided managers with the information they needed to command and control those below them in the hierarchy that characterized the management structures of industrial age and Era I information-age firms. Actually, this use of mainframes was the last gasp of this style of management: the information environment in which business operates today has simply become too complex for command-and-control management to survive. The management information systems (MIS) that mainframes provided merely prolonged the life of the hierarchical management style for another thirty years, but its eventual demise was inevitable.

As in the industrial model, the organization of MIS personnel in the mainframe computer environment was centralized and controlled hierarchically by MIS specialists who advanced through the ranks from programmer to analyst to senior analyst to project director to manager of MIS. The ranks of MIS personnel were very much like those of the military. There was a degree of uniformity in terms of the experience of the MIS personnel and the skills they were taught. Because the training of MIS staff was so specialized and required the logging in of long hours tediously programming the computer line by line, most MIS personnel did not have much experience with other aspects of their firm's activities. Because of the investment in time required to be a computer programmer or systems expert, virtually none of the non-MIS personnel of the company became involved in the operations of the firm's computers. This total split of MIS functions from the business activities of the firm created two cultures with little or no understanding of each other's functions or needs. This split was absolute until the advent of microcomputers and their integration into the work flow of virtually all organizations, from the largest to the smallest. The split also helps explain the difficulties that many microcomputer end users had with their MIS departments and vice versa. MIS personnel and end users come to computing from completely different perspectives. It is only through a process of dialogue and the integration of work groups that these two diverse groups can learn to work together.

The development of software was one aspect of MIS operations during Era I which did not conform to the model of industrial organization. Each manufacturer of a mainframe computer had its own proprietary operating system; once a firm committed to a supplier, it was locked into a unique system and could not purchase software developed by any other suppliers. In fact, for most of the information-processing jobs that a company wanted its computer to perform, software had to be created specifically for that application. With the exception of a small number of applications, software had to be crafted by a team of programmers and analysts for each application. The code that they created could not be used for other information-processing activities.

Era II computing has changed dramatically. There are only three standards for operating systems: one was developed by Microsoft (MS-DOS) for the IBM personal computer, one by Apple for the Macintosh, and the third, UNIX, which is an alternative to MS-DOS. The Macintosh system, which features a graphical user interface and a mouse for clicking on icons, is much easier for the unsophisticated end user to work with. Microsoft, the creators of MS-DOS, duplicated many of the features of the Macintosh graphical user interface with one of their own called Windows. It is quite likely, however, that the two current standards for operating systems, the Macintosh and MS-DOS/Windows, will merge because of the recent release of the IBM/Apple joint-venture product, the Power PC, which can run software from either standard.

An important feature of Era II software which distinguishes it from hand-created Era I software is that it is mass-manufactured and can be purchased shrink-wrapped directly off the shelf with no or little need for individualization or installation by a professional. It is ironic that modular software mass-produced using industrial-age techniques became the engine for Era II computer usage. End users network with co-workers, suppliers, and customers, and students working in either the classroom or at home have the freedom to customize information-processing activities in a decentralized, nonhierarchical, and fully integrated environment which fully embodies the spirit of the information age. Era II computing, which is microcomputer-based, nonhierarchical, networkable, and open, is having a profound effect on education, on the workplace, and also on the relationship between schools and work.

The Impact of Microcomputers on Education and Work

There has been a steady invasion of microcomputers into our schools and workplaces for the past fifteen years, resulting in the transformation of both domains. This new technology challenges us to evaluate the way we have organized our education system and workplace environment. We must understand the way in which the new information technologies will continue to change our schools and workplaces and require them to perform new tasks that they never before had to consider. The impact of computer systems and the social interactions they engender will therefore become a mandatory subject of study for all those concerned with industrial and/or educational planning.

As we discussed in Chapter 4, the education system is particularly sensitive to the influence of communication media and, consequently, has been shaped by them over the centuries. The origins of schools and the formal classroom style of teaching can be traced to the need in Sumer some 5,000 years ago to train young people to use the medium of writing. Many aspects of school and classroom practice have changed surprisingly little since that time. If the invention of writing was the impetus for starting schools, then the Gutenberg printing revolution represents another benchmark in the evolution of education. Low-cost printed material made possible the delivery of educational services to a mass audience. It inspired a system of schooling whose primary task was to teach children to function in the new print environment. Its goals and values, which came to be colored by a print mentality, still color today's schools.

We are now in the midst of a third major communication revolution, one created by electronic information devices. This revolution promises to have as profound an impact on society and education as the printing press had 500 years ago.

Computing devices entered the school system in the early sixties as an administrative tool in the form of mainframe computers. This technology was initially used to teach computer programming courses primarily in secondary and postsecondary schools. Mainframes were later used as delivery systems for computer-assisted instruction (CAI) in which students mastered the content of lessons introduced by their teachers through drill and practice, or were introduced to new content through tutoring. CAI was based on a behaviorist model in which the user was a passive recipient of information. CAI offered the advantage of individualizing the student's learning program and automating certain of the teacher's functions such as tracking and evaluating a student's progress (Kurland and Kurland 1987, 319–20). The advent of microcomputers, which entered the schools at the grass-roots level, decentralized and dramatically changed the nature of computer use in schools. Because of their affordability, microcomputers found their ways into virtually every school in North America, with high schools averaging forty to fifty machines per school and elementary schools about twenty machines per school (Becker 1991a, 6–9).

At first, microcomputers were used as delivery systems for CAI, but soon new applications began to appear, such as word processing and LOGO, a programming language for primary-school children, in which students could more actively control their learning experience. As computers began to proliferate, more and more computer literacy and programming became educational goals in their own right. LOGO and other programs, such as Rocky's Boots and Gertrude's Puzzle, created microworlds which elementary-school pupils could explore on their own to create their own learning experiences. The tool use of computers for functions such as word processing, spreadsheets, and databases allowed the greater integration of computers into the general curriculum. As already discussed, new applications such as hypertext and multimedia, in which images, sounds, and texts are integrated, hold out the promise for more changes in the use of computers in education.

A number of authors (Papert; Coburn; Pitschka; Bitter and Camuse; Taylor) suggest that the microcomputer will be a source of revolutionary structural change in the modern education system and in the contemporary workplace. What is the basis of this thesis? Other media introduced into the business world and the education system have had significant impacts but they have not made truly revolutionary or structural changes. Certainly, the telegraph, the telephone, and electric lighting made significant changes in how business was conducted, but not enough to affect its hierarchical structures. The impact of these inventions on education was even less. The mass media of radio, recordings, television, and movies had a greater impact on education than on work, but they were not the radical change agents they had been touted to be. Educational television is one of the more valuable forms of television programming, but it has not had a lasting impact on education or been the source of any cognitive breakthroughs. Mainframe computers had very little, if any, effect on education. Their effect in business has been significant from a quantitative point of view but not from a qualitative one. They have allowed firms to expand the scope of their operations and have automated a number of traditional business processes, but they have not changed the basic corporate hierarchical structure that has characterized companies since the beginning of capitalism and industrialism. What is different about the microcomputer and client/server systems and their power to effect structural change that distinguishes them from other media?

I will respond to this question by postulating nine significant impacts microcomputers are having (or beginning to have) and suggest why they will have a profound and liberating effect on both work and education:

1. The microcomputer is the first medium to successfully vie with television for the attention of both adults and young people since the advent of commercial television.
2. The microcomputer is a medium of communication which is interactive and hence has the potential to promote exploration and discovery.
3. The microcomputer is a medium which promotes educational activity and hence can liberate untapped potential in students and workers.
4. The microcomputer is an ideal medium for delivering and promoting individualized learning and/or individuallized work productivity.
5. The microcomputer has the potential to promote a positive attitude towards work and learning and encourage a positive self-concept.
6. The microcomputer has the potential to profoundly

change teaching patterns within the classroom, as well as the social interactions of pupils and teachers in schools and of staff and management in the workplace.

7. The microcomputer is the first educational technology to be introduced into schools at the grass-roots level by teachers, parents, and students, and into the workplace by workers and middle managers rather than by senior administrators.
8. The microcomputer is a medium through which the curriculum can be integrated in the school and the flow of work can be integrated in the business world.
9. The microcomputer will act as an agent of reform challenging the notion that hierarchical control-and-command structures are the most efficient way to conduct business or that a formal school structure is the best setting for educating children.

These nine postulates suggest that microcomputers will play an important role in the future development and reformation of business and education. If these ideas are correct, then understanding the effects of microcomputers and their associated software will be essential for the educational planning required to prepare children for the challenges of the computer age and for the planning of programs of lifelong learning and training for workers in the business world.

The following elaboration of these nine points does not claim they are empirically confirmed facts. Rather, they are hypotheses requiring further study. They have been formulated on the basis of direct observation of computer usage in the classroom (Sullivan et al. 1986) and the workplace.

Microcomputers and the Ninefold Enhancement of Education in the Schools

We are only beginning to understand the range of the applications of computers in education. The only way to study this new medium and its effects on education is through the collection and analysis of observational data from a variety of sites. This process will help us understand the issues surrounding the planning, use, and implementation of microcomputers in the school system.

The nine postulates formulated above are tested against the wealth of observational data on the classroom use of microcomputers collected in a comparative ethnographic study in which I participated (Sullivan et al. 1986). The conclusions reached in this study of the use of computer technology in elementary schools were based on classroom observational data, information gathered in interviews with teachers and pupils, a survey with the pupils, and historical data (Innis 1951, 1972; McLuhan 1962, 1964; Logan 1986a; McLuhan and Logan 1977) on the use of traditional media such as writing and print.

1. Attention Span

The microcomputer is the first medium to successfully vie with television for the attention of young people. I do not wish to overstate the case and claim that all children are attracted by this new medium or that once they are attracted to it, they never tire of it. The students' attitudes towards computers are not uniform and those who are generally enthusiastic are sometimes bored, while those who are generally bored are sometimes enthusiastic. There were instances where we observed increased attention span and concentration as a result of the technology. Thoughtful instruction or guidance is a necessary ingredient to ensure the successful use of computers in the classroom, however.

The microcomputer's capacity to engage children and capture their attention through interactive video images is a property shared with video games. This property, not shared by textbooks, can be exploited to design engaging educational software, which offers competition to the fast-paced, rapidly changing images of television. Microcomputers, however, are not an automatic draw; a well-designed book can and must compete successfully with computers. While microcomputers are a versatile educational medium, there are still many valuable educational experiences which can only be obtained with books.

2. AN INTERACTIVE MEDIUM

The microcomputer is a medium of communication which is interactive and hence has the potential to promote exploration and discovery. Television, in sharp contrast to the microcomputer, is a totally passive medium. Children watching its flickering mosaic images quickly become mesmerized. Commercial television is mental junk food which leaves the minds of its viewers unnourished and inactive, as can be detected by the glazed look in the eyes of children watching television. The microcomputer, on the other hand, if suitably programmed, can be used to stimulate an active response. When used skillfully and creatively by the classroom teacher, it can function as a medium of exploration and discovery. Moreover, the microcomputer provides the user with a sense of control both over learning and over the data or information contained within or generated with the computer.

The microcomputer is not the first interactive medium in education; the written word is certainly another example. Although the book is only a delivery vehicle, it engages the mind interactively because of the reader's need to decipher its message. Writing and reading entail the active coding and decoding of information from visual signs into language and vice versa. Alphabetic writing also requires the active analysis of spoken words into their basic phonemes so that each phoneme can be coded by a letter of the alphabet. Writing has traditionally functioned as a medium for exploration and discovery for both schoolchildren and adults.

Indeed, the creative output of science and the humanities would not have been possible without this evocative medium. "Ideas develop from interaction and dialogue . . . especially with one's own writing" (F. Smith 1982, 204).

The computer has also had an impact on stimulating the development of new ideas because it, too, is an interactive medium, on two accounts: as a medium of communication for the written word, and as a data-processing device. The computer provides the user with an additional dimension of interactivity compared with writing, namely, the capacity to manipulate data. It is the most powerful research tool and stimulus to new ideas developed since the invention of writing, numerical notations, and the printing press, and like these media its use will contribute to the further cognitive development of its users.

The microcomputer is an interactive medium that pupils are able to operate in an exploration-and-discovery mode. But the microcomputer by itself is not necessarily interactive; it must be programmed in a particular way to realize this potential. Even with an interactive form of software like LOGO, however, its use does not necessarily lead to exploration and discovery on the part of the pupil. Students still require the guidance of a teacher to integrate their use of the microcomputer with the rest of the educational program and to show them how to use the software in a way that will promote self-growth.

3. AN EDUCATIONAL MEDIUM

The microcomputer is a medium which promotes educational activity and hence can liberate untapped potential in students. We observed in most cases at our six sites modest educational gains with word processing, drill and practice, and programming. Particularly impressive were the results achieved in the area of special education, where the computer supported all aspects of the teacher's program and students working independently on the computer increased both their computer and academic skills (Sullivan et al. 1986, chap. 8).

Because it is interactive, the microcomputer allows users to develop new cognitive skills as well as reinforce their literacy skills. Television, on the other hand, does not promote education because it is a basically non-interactive medium which merely transmits information to passive viewers. Television informs but it does not educate. Television viewers do not develop new cognitive skills as a result of using the medium. This helps explain why television was such a disappointment as an educational medium. Information does not equal education; the ability to process information is an essential part of education and television does not promote this activity.

The fact that television allows children to become better informed has actually had certain negative consequences. From a very early age, youngsters in the television age are exposed to almost all aspects of life through the medium of television, and as a consequence are "gray at the age of three," to quote McLuhan (McLuhan 1964). This may explain why the present generation of children seem to know so much yet understand so little. They are already jaded about life so that their teachers have little new information to offer them and as a consequence find it difficult to hold their attention. Children in the television age also become intellectually lazy since no effort is required to absorb TV-transmitted information, in comparison with learning from written material.

Print promotes educational activity because in addition to informing its readers through its content, its use, unlike that of television, requires that the individual possess certain cognitive skills to decode its content. To use the alphabet (Logan 1986), a writer (or a reader) must first learn to analyze each spoken word into its basic phonemes and then code for writing (or decode for reading) those phonemes into (or from) meaningless visual signs, the letters of the alphabet. In addition to promoting the skills of analysis and coding, alphabetic literacy teaches abstraction and classification. Computer literacy involves similar skills of analysis, coding, abstraction, and classification, and hence reinforces the skills of literacy and the left-brain modes of organization associated with them. Moreover, computers promote right-brain forms of patterning and organization because of their use of more global categories such as files, fields, and databases. This is one of the reasons that the microcomputer has the potential to become such a powerful teaching tool.

Television, on the other hand, has none of these features; this is why it can only inform and cannot really educate. Education can be defined as what is left over after you have forgotten all that you learned. This definition is facetious, but it contains an important insight into the effects of interactive media such as the written word or the computer. Learning facts is not an education, but learning how to process information or data is. A possible reason that higher learning or science did not develop in preliterate societies is that there was no interactive medium available to store and process information. Writing provided such a medium and print reinforced or enhanced this quality. This explains the sudden burst in education, scholarship, research, and science, first after the invention of writing in Sumer and Egypt, then after the development of the phonetic alphabet in the Levant and Greece and, finally, after the invention of the Gutenberg printing press in Renaissance Europe (Logan 1986a). We are experiencing a similar explosion of activity with computers.

The microcomputer combines the educational advantages of the written word with the graphic capabilities of video. It is a multisensory medium capable of providing the same information in alternative modes (textual, graphic, or animated). Computers can deliver more data to students in easily accessible forms, and, if programmed cybernetically, only as required.

4. INDIVIDUAL LEARNING

The computer is an ideal medium for delivering and promoting individualized learning. At all of the sites we observed, children used their computers to a certain extent in an individualized mode. A factor that contributed to the success of the individualized instruction at one site was the considerable talent the teacher had developed for juggling the needs of five or six students by answering their questions and encouraging their work as she moved from one workstation to another. We found less success if the teacher was not as skilled in this way. Care must be taken to make sure pupils are properly supported when they are working at the computer in an individualized mode. The microcomputer supports individualized instruction but it is not an automatic teaching machine.

One of the advantages of microcomputers for individualized instruction is their ability to provide instant feedback without embarrassing the user if a mistake has been made. Conversely, children derive a great deal of pleasure from sharing with their classmates the positive feedback they receive from the computer.

The use of microcomputers in education reminds us of the importance of individualized learning, not only with computers but with other media, as well. Individualized learning is not unique to computers. The book is also a medium that is conducive to individualized learning. Before the advent of printing, books were a precious handcrafted commodity chained to tables and studied by individual students rather than by a class of students at once. Before printing, each student studied the manuscript most appropriate for his level of knowledge. The printing press and the mass production of textbooks changed the pattern of book usage in school. The self-taught person who does not attend school but learns by reading books continues to pursue an individualized style of learning. Schools, however, where an entire class studies the same printed textbook and is expected to learn their lessons at the same time, no longer provide individualized instruction. Textbooks in modern schools are used as a broadcast medium which students must follow in a lockstep program of learning. It is up to teachers to individualize the instruction of their pupils. It takes more time and effort on their part, but it is worth it when one considers that each child has his or her own special needs and cannot be treated as an indistinguishable unit on an assembly line.

Even though computers invite an individualized approach to learning, there is still the danger that they, too, can be used in the broadcast mode. At present, this danger is not grave because there are still relatively few computers in the schools, but that will not always be the case. Computers should not be used the way textbooks were, to coordinate the studies of all pupils in the same classroom. Indeed, computers can be used as dynamic tools to individualize the use of text-based information.

5. THE AFFECTIVE DOMAIN

The microcomputer has the potential to promote a positive attitude towards learning and encourage a positive self-concept. Microcomputers have enjoyed wide acceptance and enthusiastic responses from schoolchildren worldwide across all socioeconomic classes, grade levels, and ranges of aptitude (intelligent, normal, remedial, learning-disabled, physically challenged, and so on). There is, however, a small but significant percentage of children in every group who are either hostile or indifferent to the medium. But most pupils like computers because of the way this technology captures their interest and imagination.

The enthusiasm for microcomputers often translates into a positive attitude towards schoolwork and learning in general. This is not to suggest that every child who uses a computer automatically begins to enjoy school. Computers are not a panacea, but they do provide a significant number of children who may be frustrated by the predominantly print-based education program of our education system with an alternative mode of expression and learning.

6. SOCIAL AND TEACHING INTERACTIONS

The microcomputer has the potential to profoundly change teaching patterns within the classroom, as well as the social interactions of pupils and teachers. One of the earliest fears surrounding the use of computers was the concern that schoolchildren would become antisocial automata, so transfixed by their interactions with computers that they would lose interest in interacting with one another. The intense social interaction and cooperation which surround computer use in classrooms (Sullivan et al. 1986) show that these fears were ungrounded, illustrating once again that the effects of media are sometimes counterintuitive. The microcomputer actually promotes cooperation and social interaction.

Whenever schoolchildren are using computers, either in small numbers (one or two per class) or in a lab with many machines, their social interactions are very intense. The students are constantly conversing with one another, answering one another's questions, and discussing what they are doing. They often verbalize and describe their exploits on the computer even when no one is there to listen to them.

Books do not invite the same level of intense social interaction. There is something intrinsically private about reading. Silence is requested in a library but not in a computer laboratory; mandatory silence would kill the social interactions necessary for the successful use of computers. People use books in public places such as trains or airports to achieve privacy; one hesitates to interrupt someone who is reading or writing (either manually or with a typewriter), but one feels less compunction about talking to someone working at a computer. Somehow, it is not as disruptive to distract a computer operator as it is a typist. If you cause the typist to make a mistake, it is a real intrusion because of the special effort required to correct errors, whereas with word processing little harm can be done because of the ease of correction.

On the whole, our observations (Sullivan et al. 1986) confirm that the social interactions of students increase when they use computers, whether in a classroom or laboratory setting. The teachers at the various sites confirmed this. One claimed that young people are more social at computers, which she thought was beneficial if properly controlled. Another found the increased social interaction disruptive. "It is more difficult keeping students on task and getting them to organize their time" (Sullivan et al. 1986, chap. 7). Another teacher agreed. "The noise that the children make using the computers was distracting to the other children. Sometimes, a whole lot of children went up suddenly to look at the computer, and the teacher would have to yell at them from the back of the room to sit down" (Sullivan et al. 1986, chap. 6).

Data from our interviews and surveys with the schoolchildren at the sites also confirm the hypothesis that their social interactions around the use of the computer are intense. The children indicated overwhelmingly that they preferred working with others rather than alone because they could help one another. The loners preferred the absence of interference. Most of the children reported that there were times when other students interfered with their work on the computer without being asked, which they found to be a nuisance. They also indicated that some students copied their work, but they did not seem to mind this. They indicated that they enjoyed talking about their computer activities with other children, although some found that it distracted them from their work. The children said they spent a good deal of time helping one another and felt that many of their classmates knew a lot about computers.

The most constructive social interaction around the use of microcomputers in the classroom is peer teaching. Many teachers take advantage of this phenomenon. If they wish to teach a new technique to their class, they often show it to the three or four computer whizzes in their class; within two to three weeks, the entire class will know the new technique. The knowledge has magically diffused through the whole class as a result of peer interactions. One school board encourages its teachers to take advantage of peer instruction and even provides guidelines for identifying "student facilitators."

Peer teaching completely restructures social and pedagogical patterns in the classroom. Instead of all learning interactions radiating, often in an impersonal manner, from teacher to pupils, a much richer network of interactions emerges. The teacher no longer possesses a monopoly of knowledge, because the microcomputer democratizes expertise. Frequently, one or two experts who know more than the teacher about certain aspects of computers come to the fore. These students alter the social structure of the classroom. The phenomenon of student experts is not a transient one which will disappear once teachers learn to use computers. Because the technology will continue to evolve and improve rapidly, it is likely that the present generation gap between teachers and their computer whizzes will continue for many years.

Peer teaching, the presence of student experts, and the intense socialization surrounding the use of computers all contribute to a completely different teaching style. Although the role of teachers has changed, their importance has not been diminished. The success of computer activities in classrooms depends heavily on the teacher.

7. A GRASS-ROOTS MOVEMENT

The microcomputer is the first educational technology introduced into the school system at the grass-roots level by teachers, parents, and pupils rather than by administrators. Microcomputers were not introduced into the classroom as the result of decisions taken by educational planners or administrators, as was the case with educational television. Teachers who owned their own personal computers found them to be useful tools. In many cases, they were first used for administrative purposes by teachers, then by students for their schoolwork.

The grass-roots response of pupils and teachers is the key to the success of microcomputers in schools. Suddenly, an alternative learning environment came into existence which suited the learning styles of a much greater variety of schoolchildren. Microcomputers create new opportunities for those who are at a disadvantage operating in a purely print environment, and hence further democratize educational opportunities by providing an alternative medium to print.

One of the factors that paved the way for the use of microcomputers in the school system was how the organization of primary-level classrooms changed after the introduction of television. The shift from desks bolted to the floor at regular intervals in neat rows to ones that could be moved around was the start of a revolution in education. Teachers began to react subliminally to the competition of television by introducing activity centers into their classrooms to more effectively engage their pupils and break up the monotony of sitting at a desk all day long. The use of audiovisual aids encouraged this development even more. By the time microcomputers were introduced to the classroom, the stage had been set for their use.

One of the important and interesting effects of the use of microcomputers is the autonomy it provides a teacher. The traditional classroom where most learning aids are print-based and used in a nonindividualized manner has been completely controlled by centralized authorities such as school boards or ministries of education through the choice of textbooks (using instruments such as Ontario's Circular 13) and by setting curriculum guidelines. The only options open to teachers have been those regarding teaching style and delivery. With microcomputers, the teacher has more options and greater control. The same is true for students because of the greater opportunity this technology affords to individualized learning. So, the microcomputer democratizes and decentralizes authority by giving users greater control over their activities whether they are learning or teaching.

One danger that the grass-roots movement faces is its own success. Once it is perceived that microcomputers are here to stay (and we are not far from that point), there will be a natural response from bureaucratic structures, such as school boards, to control the phenomenon, as was the case with textbooks. This impulse is dictated by the logic of bureaucratic organizations and their need to rationalize and explain the large expenditure that microcomputer-based education requires. The only way to avoid unnecessary bureaucratic interference is to create meaningful communication between administrators and their clients, the teachers, parents, and pupils at the grass-roots level of education. Breaking the lines of communication, even for the shortest time, will be disastrous. The imposition of standards, equipment, or software on teachers without the teachers' support will be counter-productive.

One of the reasons for the grass-roots acceptance of and enthusiasm for the microcomputer has been its human scale. Remember, before the introduction of microcomputers, mainframe computers had only a marginal effect on education. By contrast, students are able to conceptually grapple with micros. They can "get their arms around the computer," so to speak. They are able to see all of the components of the machine at once. The operation of the disk drive is familiar to them because of the parallel of the floppy disk with the CD disk and the tape cassette. The ability to control and be aware of all the inputs is a great advantage to students first learning how to use a computer. Indeed, it provides them with a sense of pride and independence.

8. INTEGRATING THE CURRICULUM

The microcomputer is a medium through which the school curriculum can be integrated. One of McLuhan's basic insights into the nature of electronic media was that they tend to encourage the integration of information rather than its fragmentation, as is the case with print. The current school curriculum, a product of the print era, is organized into separate subjects which are, more often than not, taught independently. The spirit of specialism can be found throughout the education system from the primary grades to postgraduate studies. The arrival of the computer has done very little to reverse this trend. For McLuhan, however, the integration of the curriculum is inevitable. "In education the conventional division of the curriculum into subjects is already as outdated as the medieval trivium and quadrivium after the Renaissance. Any subject taken in depth at once relates to other subjects" (McLuhan 1964, 347). Note that this thought was expressed thirty-one years ago, long before the advent of microcomputers and the use of computers in schools.

Word processing is one computer application which integrates English language and communication skills with other subjects such as the natural sciences, history, geography, and other social studies. Other applications such as spreadsheets and databases have shown preliminary promise for the integration of mathematics, science, and social studies. While the level of integration is still quite modest, the potential for the future is unlimited and will be an important source of reform in the education system.

9. ALTERNATIVE FORMS OF EDUCATION

The microcomputer will act as an agent of reform challenging the notion that a formal school is the best setting for educating children. This fact was recognized by D. Mersich as early as 1982: "Advances in computer-aided learning (CAL), together with the discretionary attendance provisions of the law, will soon make it possible for private enterprise to go into competition with the public education system, whip it in quality, and make a profit at the same time" (Mersich 1982, 37).

The competition that private entrepreneurs might give to the school system will not force public education out of business. It might siphon off some of the best teachers, however. We can only hope the competition will force the publicly supported education system to reform some of its practices to better meet the needs of its clientele. This is an extremely healthy prospect for contemporary education and should have an effect similar to the one the printing press had on the reform of medieval universities and the preparatory schools that fed them. The printing press updated and revitalized the education system of its day. The microcomputer will have a similar effect with proper implementation in the school system.

The claim that the microcomputer has the potential to revitalize education cannot be empirically validated because the presence of microcomputers in the school system is such a new phenomenon. There were a number of promising features of computer usage in the six primary-school sites we observed, perhaps foreshadowing the type of reform and renewal of the school system that computers could help generate. While the successes that were observed were modest, they are the type of experiences that educators can build on to exploit the educational potentials of microcomputers. The fact that the special-education teacher at one site was able to use the computer as a tool to engage her pupils' attention and then use their interest to deliver an individualized program of education which stimulated their literacy skills is significant given the rudimentary level of our understanding of how to use computers effectively in schools.

The moderate success of the computer labs is another hopeful sign. Even the very modest educational gains in the classroom use of computers portends well for the future. What is important is not just the level of success we observed in the course of one year of observation, or the progress that has been made in the few years since the introduction of the first primitive microcomputers into the school system, but rather the trend line and the rate of growth of the phenomenon. There is evidence that microcomputers possess the potential to renew our education system, but that renewal will not be automatic. The question is how the implementation of microcomputers within the education system can be organized to realize that potential. This constitutes one of the major challenges facing today's teachers.

Another trend we observed which could contribute to the renewal of the school system was the strong interaction between school and home computer usage. Children with computers at home enjoyed a definite advantage. While at the moment this advantage represents an equity issue between working- and middle-class children in the short-term, the drastic reduction in the cost of computers expected in the future should cause this disparity to disappear. What is important is that computer usage represents one of the first activities in a long time that integrates schoolwork and home recreational activities. Computers could become an important tool to combat the alienation children feel between their schoolwork and what they experience outside of school. Given the increasingly important role computers will play in the world of work, school computer programs will inject the element of relevance that educators have been desperately seeking to motivate their pupils.

Microcomputers and the Ninefold Enhancement of Education in the Workplace

Today's workers and employers are under enormous pressure just to remain employed or in business. During the industrial age, workers or employers needed to learn one set of skills that would in most cases last them a lifetime. This is no longer the case. More and more, workers and businesses are finding themselves displaced by technology. Their salvation lies in ongoing education so that their skills remain relevant to the current environment of rapidly changing technology and skill requirements especially due to computers. All technologies, as noted in Chapter 2, are a source of both service and disservice. Job displacement by technology is a disservice. The increased productivity that microcomputers permit is a service. So, too, is the fact that microcomputers stimulate learning on the job and make lifelong learning more possible. Just as the microcomputer has stimulated reform of the school system, it can perform a similar task in the world of work and for basically the same reasons. The ninefold enhancement of education by the microcomputer, which I originally formulated for children in the classroom setting, also provides a good description of increased productivity and learning in the workplace. The conclusions herein are based on reports in the literature and some direct observation of computer usage in the workplace and at a computer-training center in Toronto.

1. ATTENTION SPAN

The microcomputer is the first medium to successfully vie with television for the attention of adults. Workers are by and large enthusiastic about microcomputers. There are a few exceptions, generally among individuals who have had difficulty learning how to use their computers because of poor training or no training. For workers who have mastered its use, there is satisfaction with computing for a variety of reasons:

1. Users feel they are more productive.
2. They find that their work is easier and less tedious.
3. They enjoy the fact that they are constantly learning new skills or improving their existing skills.
4. They enjoy following the ongoing and continuous development of the technology.
5. They feel less trapped in their positions because they have generic skills which can be transferred to other jobs in their firm or elsewhere.
6. They enjoy the intellectual challenge of using the technology.

As with schoolchildren, adults enjoy the graphical features of microcomputers. The graphical user interface with pull-down menus has made the use of computers simpler and more enjoyable. Indeed, there is an almost gamelike aspect to some functions. The loss of attention span applies to today's adult population as well. Graphical user interface helps users overcome weak attention spans and allows them to deal with information in small, digestible bytes. Their aversion to print is similar to that of children, and therefore computers, particularly graphically configured ones, make it easier for them to work with text-based information. This is another reason that Era II computing is so different from Era I computing. In Era I, computer-generated information was accessed by the end user through printouts and had to be read just like print. Era II computing involves much more direct, graphical contact with the data.

Another parallel with schoolchildren is the importance of proper instruction and guidance. Proper training, which includes a general understanding of the hardware system and its operating system, as well as specific knowledge of the software application, is the factor that most contributes to user satisfaction.

2. AN INTERACTIVE MEDIUM

The microcomputer is a medium of communication which is interactive and hence has the potential to promote exploration and discovery. One of the chief benefits of microcomputers and client/ server systems is that they allow end users to interact directly with data. This provides users with more potential for exploration and discovery. It also provides them with a sense of control over the data and information they are working with so that they do not feel overwhelmed by the increasingly greater amount of data. Because end users are able to interact with their data, they have a sense of ownership of it and consequently a better understanding of it.

The interactivity that microcomputers make possible fully accrues when they are being used as a tool to generate new information, as is the case when word processing a report, using a spreadsheet, or working with a relational database. New ideas develop from the interaction and dialogue that the use of computers allows, which enriches the work of the user and helps that person develop as an individual and as a worker.

3. AN EDUCATIONAL MEDIUM

The microcomputer is a medium which promotes educational activity and hence is capable of liberating untapped potential in workers. Because it is interactive, the microcomputer allows users to develop new cognitive skills as well as reinforce their literacy skills. The skills developed in learning how to use the computer are generic and apply to all forms of information processing, which is the root activity of most work today. Another benefit of using computers is that they allow workers to integrate left-brain analytic problem-solving techniques with right-brain forms of patterning and organization, just as the computer integrates textual and graphical information. The use of the microcomputer is probably the best form of education with which a manager can provide an employee, to prepare him or her to operate in today's business environment. And the beauty of this form of education is that the worker is learning these skills in the midst of generating productive work. The only time investment outside the office is one or two days every six months to allow the user to upgrade computer skills and stay abreast of this rapidly changing technology.

4. INDIVIDUAL LEARNING

The computer is an ideal medium for delivering and promoting individualized learning and/or individualized work productivity. Its ability to provide instant feedback without embarrassing the user makes it a very powerful educational tool in the hands of all employees, assuming they receive enough initial instruction to get started. The microcomputer is also an excellent medium for individualized productivity because it provides workers with an information environment in which they can work in peace and at their own pace. The microcomputer has also given rise to a huge increase in the number of people working at home in what Toffler labels the "electronic cottage" (Toffler 1980, 26, 210–23).

5. THE AFFECTIVE DOMAIN

The microcomputer has the potential to promote a positive attitude towards work and encourage a positive self-concept. Adults feel a sense of pride when they are able to create a professional-looking product. Another factor motivating workers to use computers is their belief in its importance for the advancement of their careers. The computer relieves much of the tedium of work because it does repetitive tasks so easily. It also provides stimulation and a constant source of challenge and hence makes people's work more interesting.

6. SOCIAL AND TEACHING INTERACTIONS

The microcomputer has profoundly changed the pattern of social interactions among staff and management in the workplace. It is generally recognized that the use of this technology has flattened out hierarchical structures and led to a team-oriented rather than command-and-control style of management. The intense social interaction and cooperation which surround computer use in the workplace parallel the experience in the classroom. The microcomputer promotes cooperation and social interaction because the users feel united by a common bond to overcome their perception of the machine as "stupid." The cooperation that develops between users helping one another translates into good teamwork, which carries over to other activities.

7. A GRASS-ROOTS MOVEMENT

The microcomputer is probably the first major piece of technology that was introduced into the workplace at the grass-roots level by workers and middle managers. And this much against the wishes of the MIS (management information systems) departments who ran their organizations' information operations through the use of mainframe computer systems. MIS personnel regarded microcomputers as a nuisance. For the end users and their managers, this technology was a godsend because it allowed each department to control its own computing because they saw that microcomputers could be used to get their departments' work done more efficiently. It was this grass-roots response that was the key to the initial success that microcomputers enjoyed in large organizations.

One of the reasons for the grass-roots acceptance of and enthusiasm for the microcomputer was its human scale. Before the introduction of microcomputers, mainframe computers were inaccessible to the ordinary worker (defined as someone not in MIS); even after the introduction of terminals, the end users' use of the mainframe was severely limited.

8. INTEGRATING BUSINESS PROCESSES

The microcomputer configured in a client/server architecture has the potential to provide a medium through which all the business processes of an organization can be integrated. McLuhan long ago pointed out that computers would integrate all information processes in an organization rather than fragment them, as had been the case with print. Without ever using the term, he conceived of business process reengineering over thirty years ago. "With electricity as the energizer and synchronizer, all aspects of production, consumption, and organization become incidental to communications. The very idea of communication as interplay is inherent in the electrical, which combines both energy and information in its intensive manifold" (McLuhan 1964, 307).

9. ALTERNATIVE FORMS OF EDUCATION

The microcomputer will act as an agent of reform, challenging the notion that a hierarchical control-and-command structure of management is the most efficient way of conducting business. This is not to suggest that management will be replaced by computers. Managers will be needed to plan the overall direction of an organization, define its objectives, and develop its strategies for achieving those objectives. But management will no longer need to bother with the day-to-day tactics of running an organization. Client/server systems with appropriate communication links, information-processing tools, and adequate information will allow tactical decisions to be taken by employees. The managers' job will be to set up the tools and to provide the training. The client/server system becomes the "virtual manager," a tool that allows its users to self-organize. This is the radical nature of reform that computing will eventually bring to the workplace.

The printing press updated and revolutionized the world of commerce in its day. The microcomputer will have a similar effect. A number of promising features of computer usage in the world of work foreshadow the kind of reform and renewal of commerce that computers could bring. These features involve the transformation of today's firms into "learning organizations," as has been suggested by Peter Senge (Senge 1993), and the transformation of workers into "lifelong learners."

6

LEARNING A LIVING IN THE WIRED WORLD

Schools can no longer perform the whole job of providing students with all the tools and knowledge they need for their careers. Technology is changing too fast for that. Today, the best the schools can do is to prepare students for the task of keeping themselves continuously updated.

Given these facts, every employer must plan some form of in-service education; otherwise, their employees will fall hopelessly behind. In addition to staying up to date with technology, employers should also be actively concerned with their employees' educational development, given that most workers are now or soon will be "knowledge workers" who have to deal with information all day long. The operation of a business or any other mission-oriented organization therefore requires an educational strategy or plan as urgently as it needs a marketing or financial plan.

The job of a company's human resources department entails the constant upgrading of employees' skills as well as their general intellectual development. And each employee who wants to advance must make a plan for his or her own personal development through a program of lifelong learning. An educational program is essential for survival in the business world for both the organization and the individual.

This chapter will help readers develop a plan for their own development or for their organization. It attempts to answer three of the questions posed in Chapter 1:

1. How is computing changing work patterns and the organization of social institutions?
2. How can work be organized to naturally promote learning?
3. How can education in the workplace be better organized to improve productivity so that learning becomes a lifelong activity and workers are properly trained to do their jobs?

From Global Village to the Wired World

> Real, total war has become information war. It is being fought by subtle electric informational media.
> (McLuhan 1967, 138)

Despite the fact that computers have begun to alter our notions of language, communications, culture, education, commerce, and government, most people are largely unaware of the profound impact of technology and carry out their activities largely unaffected by the changes surrounding them. This is not altogether surprising; those who lived through the Gutenberg revolution were equally unaware of the major shifts brought about by the printing press. The pattern of unawareness repeats itself each time a new technology is introduced. As McLuhan wrote, "Every technology contrived and outered [sic] by man has the power to numb human awareness during the period of its first interiorization" (McLuhan 1962, 187). We are currently passing through such a period. The rate of change, however, is so great that we can no longer afford to ignore the process of change. The Gutenberg revolution stretched over centuries; the time scale of the computer revolution is only a matter of years. Nevertheless, in this very short period of time, our institutions are falling out of synch with contemporary economic and social realities.

That we live in an information age is a twentieth-century cliché destined to become a twenty-first-century reality and archetype. The planet *has* shrunk to the dimensions of a global village. All of the institutions of our postindustrial, postcapitalistic society, including the family, the economy, corporations, small businesses, primary and secondary schools, universities, local and national governments, the civil service, political parties, and the media are undergoing rapid and profound structural changes. As a result of these changes, we suffer from a number of information-age maladies, including future shock, information overload, political gridlock, resurgent nationalism, and ethnic xenophobia. These are the symptoms of the "information war" shell shock and the post-traumatic stress with which we struggle daily.

We are desperately trying to establish a balance between the change represented by the hundreds of technological opportunities that the future holds and the stability and comfort of our traditional institutions. Our collective response to the challenges of the information revolution has taken the interesting form of both moving forward into a new era, a new millennium, by embracing the use of computers and other forms of the new information technology, and at the same time attempting to return to some of the values of our past by reengineering our businesses, restructuring our government and its social assistance programs, reforming our education system, and renewing our spirituality. The key idea in all the schemes of renewal, reform, reengineering, and restructuring is that the relentless flow of information cannot and must not be stopped or even slowed down, but that it must be controlled and harnessed by the use of information-technology hardware, advanced software systems, expert systems, cybernetics, general systems theory, and artificial intelligence. In fact, plans are being made to increase the flow of information through the construction of the "information superhighway" which will merge computer, LANs, WANs, e-mail, Internet, telephone, fax, cable, television, video, radio, film, smart cards, ATMs, and electronic banking services. Our planet will be encircled by one continuous grid or network of information connecting homes, business, banks, schools, government offices, and information providers. The global village is becoming the wired world.

The standard approaches to the "information crisis" — that is, those approaches that focus exclusively on the technology and the flow of information — can only offer partial solutions to the challenges of living in a wired world. The problem we face is neither technical nor quantitative, but qualitative. The challenge we face is not that the amount of data and information has exploded exponentially, but rather that we must deal with new patterns of communication and information processing and the impacts they are having on institutions that were designed for different methods of data handling. A quantitative change creates a qualitative change. I believe that the so-called information crisis is not due to the rapid increase in the amount of data that is currently being processed, nor is it due to the speedup of transactions or any of the technical challenges required to deal with this increased and accelerated flow of information. It is due to our lack of the "soft" skills that are required to integrate the new techniques of data handling and information processing with the organizational structures of the social, political, economic, and cultural institutions of our society.

Our ability to handle vast amounts of information quickly and economically is prodigious and growing rapidly. It is the transformation of our society and its institutions to accommodate the new wired reality which is lagging behind. Institutions designed to handle a much slower and smaller flow of data and based purely on paper forms can no longer cope with the information explosion, and hence the feeling of information overload. It is incumbent upon us, therefore, to redesign and reengineer our institutions to cope with the increased information demands of the information age. Moreover, radical changes in the way individuals are trained to work with information must be made. The older forms of education and training simply do not work and must be reengineered.

The solution to the problem we face is not technical, but social, and of the two, social problems are always more difficult to solve. The problem lies in the conflict between the techniques, processes, and attitudes that are in place to deal with the older forms of literary and paper-based information and the reality of the newer electronic forms of information that are swamping today's institutions. The current mismatch between the newer forms of information processing and the institutions designed for older forms is not historically unique. In every age and in every society that has undergone significant technological development and change, a mismatch between the new techniques and the old institutions has been a problem. Our era is no different except that the transition between the old technologies and the new ones has been more rapid and hence the mismatch has been greater.

Why Our Industrial-Age Institutions Are Not Working

Our computer skills have evolved faster than our organizational skills. Institutions organized to address the challenges of the industrial age cannot cope with the vastly increased and accelerated flow of information. The force driving the transition from the industrial age to the information age is computing and other forms of information-processing technology. They are producing what Toffler (1980) called "the third wave of human change." If our institutions are to remain relevant and serve the needs of society, they must make better use of these communications and information-processing technologies. Most modern institutions such as the nation-state, the civil service, the corporation, the labor union, the free-market system, the banks, and the education system, however, developed during the Industrial Revolution or the "second wave" of human change when mass production was inaugurated. Many of our institutions such as the stand-alone school, the church, the city, can be traced to an earlier date. They began with the advent of writing during the agricultural revolution or the "first wave." They have undergone the same bureaucratization as those institutions that originated in the industrial age. In fact, all of our bureaucratic organizations are products of the type of organization that swept in with industrialization.

If the goals and objectives of today's institutions were shaped by the needs of an economy based on mass production, their breakdown, with the dominance of informatics as the principal driving force of our economy, should come as no surprise. The industrial model which stressed uniformity, fragmentation, and specialization has been replaced by a cybernetic model which stresses diversity, integration, and the generalist approach of systems theory. McLuhan attributed the transformation to the transition from a print-based communication system to one dominated by electricity and electronics. McLuhan's analysis, developed in the early days of computing before microcomputers were conceived, foreshadowed many of the themes that have been developed in the nineties by the business analysts discussed later in this chapter. McLuhan's ideas read in light of today's developments seem to have anticipated Toffler's observation of the crucial new role of knowledge in the production of wealth (Toffler 1980, 1990), the need for Hammer and Champy's technique of business process reengineering

(Hammer and Champy 1993), Tapscott and Caston's notion of the extended enterprise (Tapscott and Caston 1993), Senge's idea of the learning organization (Senge 1993), and Drucker's ideas of postcapitalistic society and the demise of the dominance of the nation-state (Drucker 1993).

I realize that as a friend and former colleague of McLuhan's, I might be subject to a bit of bias. Moreover, it is always easy to read predictions into people's work after the events have come to pass. Having duly warned readers, I will support my position by quoting liberally from McLuhan so that readers can examine his exact words and form their own opinions. The quotes I am using all come from the introduction to *Understanding Media* and from the last chapter, entitled "Automation — Learning a Living" (McLuhan 1964).

The fact that the ideas referred to above required thirty years to germinate and required rediscovery by other authors can be attributed to four factors:

1. McLuhan's prose style was cryptic and difficult for many to follow, or to put it more charitably, he wrote like a poet and business people do not generally look to poetry as a source of inspiration and direction.
2. McLuhan had no business experience and was not considered to have any particular insight into the nature of business. As a scholar with a literary bent, his ideas were largely ignored by the business community with the exception of creative professionals in the advertising field.
3. McLuhan was not taken seriously by business people, most academics, and even many of the writers referred to who advocate interdisciplinary approaches. Specialism dies hard even for information-age interdisciplinarians. They were unable to embrace notions like McLuhan's idea that "the social and educational patterns latent in

automation are those of self-employment and artistic autonomy" (McLuhan 1964, 359). The elevation of McLuhan to the position of patron saint by the editors of *Wired* magazine suggests that perhaps a new generation, the generation of McLuhan's grandchildren, is ready to accept his mantle.

4. Finally, McLuhan's ideas were truly ahead of their time and not enough collective experience with information technology and, in fact, none with microcomputers, had been logged in at the time of the publication of *Understanding Media* in 1964. Now that millions have experienced what McLuhan wrote about fifteen years before the introduction of microcomputers, his ideas, properly contextualized, have gained popular acceptance, most often without proper attribution.

Knowledge, the New Source of Wealth

In *The Third Wave* (Toffler 1980) and *Powershift* (1990), Toffler documents how our economy has undergone a radical transformation due to information technology. Manufacturing and distribution activities once dominated the economy. It is now the information and service sectors which comprise the majority of activity. Furthermore, the information and service sectors are merging. In describing his company, the founder of one of Europe's largest computer manufacturers remarked, "We are a service company just like a barbershop" (Toffler 1990, 74). All sectors of the economy are affected by computing. "Farmers now use computers to calculate grain feeds, steelworkers monitor consoles and video screens . . . A United Parcel Service driver handles boxes and packages, drives a van, but now also operates a computer" (Toffler 1990, 74–75). Auto mechanics at Ford dealerships are using a computer-based expert system. All of these observations led Toffler to the conclusion that knowledge will play a "crucial new role . . . in the production of wealth. Instead of metal or paper money, electronic information becomes the true medium of exchange" (Toffler 1990, 238). The economist Peter Drucker also came to a similar conclusion: "The basic economic resource — 'the means of production,' to use the economist's term — is no longer capital, nor natural resources (the economist's 'land'), nor 'labor.' It is and will be knowledge" (Drucker 1993, 8). McLuhan published this idea twenty-six years before Toffler and twenty-nine before Drucker when he wrote: "The very same process of automation that causes a withdrawal of the present work force from industry causes learning itself to become the principal kind of production and consumption . . . The peculiar and abstract manipulation of information [is] a means of creating wealth" (McLuhan 1964, 351, 354). He also captured the idea of the connection between wealth and information in the following quote which also calls for the type of restructuring of organizations known as business process reengineering (BPR): "Wealth and work become information factors, and totally new structures are needed to run a business or relate it to social needs and markets" (McLuhan 1964, 357).

BUSINESS PROCESS REENGINEERING

The first use that was made of computers was to automate the fragmented business processes that were inherited from the previous era of mechanization. This approach does not take full advantage of the technology. Business process reengineering is the technique whereby the processes by which an organization conducts its business are reorganized and reengineered to take full advantage of the information handling that computers make possible. The use of computers to automate the old processes illustrates McLuhan's adage that the first content of a new medium is always an older medium. "The content of this new environment [the electronic age] is the old mechanized environment of the industrial age" (McLuhan 1964, vii). Business process reengineering reverses this phenomenon and finds a new content which is more naturally suited to the patterns of the new technology. According to the management analysts Michael Hammer and James Champy, "The real power of technology is not that it can make the old processes work better, but that it enables organizations to break old rules and create new ways of working — that is, to reengineer" (Hammer and Champy 1993, 90).

McLuhan predicted in 1964 that all industries would have to go through a reengineering exercise because the effects of electric speedup would be to integrate the processes that industrial mechanization had fragmented and create the need for new types of organization and talent. He predicted that firms would have "to rethink through function by function," their place in the economy. In a later publication, teaming up with Barrington Nevitt, McLuhan formulated the dilemma of trying to speed up mechanical processes in the following way:

> The United States moves into a period of 'inefficiency' and breakdown resulting from the application of electric speeds to mechanical processes. The computer-system specialist, in practice, is entirely a product of old literate and mechanical training. He thinks of electric technology as a quantitative increase in the performance potential of the old mechanized system. In practice, what occurs is that the old system is simply shaken to pieces by the 'vibes' to which it is subjected. (McLuhan and Nevitt 1972, 220–1)

Hammer and Champy came to the same conclusions as McLuhan through their study of management and economics:

> Reengineering is about reversing the Industrial Revolution. Reengineering rejects the assumptions inherent in Adam Smith's industrial paradigm — the division of labor, economies of scale, hierarchical control, and all the other

> appurtenances of an early-stage developing economy. Reengineering is the search for new models of organizing work . . . Task-oriented jobs in today's world of customers, competition, and change are obsolete. Instead, companies must organize work around process. (Hammer and Champy 1993, 27–28, 49)

McLuhan, too, had identified the importance of process, but much earlier: "Electricity . . . gives primacy to process, whether in making or learning" (McLuhan 1964, 347). Hammer and Champy arrive at a similar conclusion: "It is not products but the processes that create products that bring companies long term success" (Hammer and Champy 1993, 25).

The impetus for businesses to reengineer their operations is not just to obtain the most efficient use of computers. They are doing it because they are pitted in a struggle for survival and in most cases must restructure or perish. Hammer and Champy have identified three factors that are driving this process, which they call the three Cs — Customers, Competition, and Change (Hammer and Champy 1993, 17). Customers have taken charge; it is a buyer's market. Customers now have access to so much information and so much choice that they are forcing business to respond to their needs. This phenomenon was also predicted by McLuhan: "The consumer becomes producer in the automation circuit . . . Marketing and consumption tend to become one with learning, enlightenment, and the intake of information" (McLuhan 1964, 349–50). Toffler also developed a similar idea in 1980 in *The Third Wave*. During the first wave, the consumer and the producer were identical: people produced for their own consumption; then during the second wave, a wedge was driven between the producer and the consumer. During the third wave, however, the "prosumer" reemerges. There is "a progressive blurring of the line that separates producer from consumer" (Toffler 1980, 284). They are "reunited in the cycle of wealth creation, with the customer contributing not just money but market and design information vital for the production process" (Toffler 1990, 239).

The second factor driving restructuring is competition. Consumers or prosumers armed with information technology can do many things for themselves like design their own business forms or create their own newsletters. So businesses must compete with their own customers. Moreover, the number of competitors has grown at the same speed with which the world has shrunk to the dimensions of a global village. The number of people and organizations that can compete is much greater than was the case when business was conducted on a local level. Start-up companies are another source of competition. They carry no organizational baggage and "do not play by the rules" (Hammer and Champy 1993, 22). McLuhan warned of the ease with which new enterprises could be started: "The separation of power and process obtains in automated industry, or in cybernation. The electric energy can be applied indifferently and quickly to many kinds of tasks" (McLuhan 1964, 350).

The third factor driving reengineering is constant change, which Hammer and Champy suggest is a change in the nature of change itself, a point that was also made much earlier, in 1970, by Toffler in *Future Shock*. The very first author to sound the alarm of the accelerated pace of change was McLuhan, who, as noted in Chapter 2, introduced the notion of electric speedup.

POSTCAPITALIST SOCIETY

In addition to increasing the level of competition that individual firms face, the globalization of trade has also undercut the importance of the nation-state, according to Peter Drucker. "The sovereign nation-state has steadily been losing its position as the sole organ of power. Internally, developed countries are fast becoming pluralistic societies of organizations. Externally, some government functions are becoming transnational, others regional . . . and others are being tribalized" (Drucker 1993, 10–11). This latest development in the geopolitical sphere was predicted by

McLuhan based solely on his understanding of communications. "Electric speed requires organic structuring of the global economy quite as much as early mechanization by print and by road led to the acceptance of national unity" (McLuhan 1964, 353). McLuhan pointed out that nationalism "wiped out many of the local regions and loyalties" because of the speedup of information due to the printing press. In a similar manner, he suggested, the speedup of information flow due to electricity would undercut national organization in favor of global and continental associations such as the United Nations, European Union, and NAFTA. There has also been a return to and revitalization of local and regional loyalties. The decentralizing effects of electricity predicted by McLuhan and the flip back to tribal or oral structures are surfacing so that many people now embrace the credo: "Think globally, act locally." It is the national institutions that are being caught in this squeeze play. This is not to suggest that the nation-state is about to disappear, but only that it will no longer dominate the geopolitical and economic systems the way it once did.

McLuhan also saw that the electric speedup of information would neutralize the great ideological conflicts that were the basis of the cold war. "The electric changes associated with automation have nothing to do with ideologies or social programs" (McLuhan 1964, 352). The end of communism and the restructuring of capitalism have had nothing to do with the years of military and propaganda posturing of the Eastern and Western blocs led respectively by the former USSR and the United States. The transformation took place as a result of economic and social forces driven largely by information technology. There was a reason the Communists banned private ownership of the personal computer. They recognized that the electric flow of information was the one thing that their centralized hierarchical command-and-control system could not defend against and in the end it became impossible for them to do so. Their system was inundated by information and collapsed under its weight. The coup de grace, ironically, was actually due to Gorbachev's introduction of *glasnost* (openness) and *perestroika* (restructuring), which finally brought down the whole system. A similar but less dramatic transformation has also changed the face of capitalism as old-style forms of command-and-control hierarchical corporate structures give way due to the Western form of *glasnost* and *perestroika*, namely, open nonhierarchical computer-based communications and flow of information, and business process reengineering. McLuhan's predictions came to pass as Drucker proclaims in his latest book, *The Post-Capitalist Society*: "That the new society will be both a non-socialist and a post-capitalist society is practically certain. And it is certain also that its primary resource will be knowledge . . . Instead of capitalist and proletarians, the classes of the post-capitalist society are knowledge workers and service workers" (Drucker 1993, 4, 6).

PARADIGM SHIFT

In their book *Paradigm Shift*, Tapscott and Caston discuss the transformation of business operations due to the interconnectivity that networked computing environments make possible. They describe three spheres of networked operations within a business organization which are replacing the departmentalized fragmented corporate structures that characterized industrial-age firms. They are: (1) the high-performance work group, (2) the integrated organization, and (3) the extended enterprise, which is the integrated organization together with its network of associates including its suppliers, customers, distribution channels, and strategic allies or partners.

The high-performance work group is an interdisciplinary team working together within the organization through a network of computers which "provides personal and work-group tools, information, and capabilities to directly support all categories of people in the information sector of the economy . . . Enterprise architectures provide the backbone for the new open, networked enterprise" (Tapscott and Caston 1993, 15, 17).

The integrated organization consists of work groups that are able to communicate and exchange information with each other so that the organization operates in an organic and purposeful manner.

The extended enterprise consists of the integrated organization together with its network of associates with which it is doing business, whether they are customers, suppliers, allies, or partners. The extended enterprise represents a group of organizations linked by one continuous stream of information flowing between them. This integration maximizes the efficiency of the business transacted between the associates of the extended enterprise.

The Tapscott and Caston model is based on the integration of automation and communication and the need for people working together to coordinate their activities. McLuhan's insight into the integrative nature of electricity foreshadows many of Tapscott and Caston's ideas: "Total interdependence is the starting fact . . . With electricity as energizer and synchronizer, all aspects of production, consumption, and organization become incidental to communications" (McLuhan 1964, 354, 359). Without understanding any of the details, McLuhan developed the notion of the extended enterprise based on his understanding of the effects of electric information. "Naturally, when electric technology comes into play, the utmost variety and extent of operations in industry and society quickly assume a unified posture . . . Many people have begun to look on the whole of society as a single unified machine for creating wealth" (McLuhan 1964, 348).

THE LEARNING ORGANIZATION

There is one point on which all of the models of today's economy agree: that knowledge and information are the keys to the production of wealth. Combine this idea with Tapscott and Caston's notion of high-performance work groups and the integrated organization, and one has derived the notion of the "learning organization" which forms the central theme of management analyst Peter Senge's provocative book, *The Fifth Discipline*. Senge, quoting *Fortune* magazine, wrote, "Forget your tired old ideas about leadership. The most successful corporation of the 1990s will be something called a learning organization" (Senge 1993, 4). A learning organization is one that makes a commitment to team learning, continual lifelong education, and a general systems approach to its operations. These are themes that McLuhan began to talk about in the sixties. He observed that the electric speedup of information meant that a single set of skills would not last a worker through his or her lifetime. The only solution to this dilemma is lifelong learning and the acquisition of generic skills rather than specialized skills. Just as electricity separates the source of its power from the processes to which it directs its energy and is thereby flexible and easily directed to different tasks, so too must today's worker possess the same flexibility. The acquisition of generic skills is the only assurance or security that an individual has in an economy that moves like quicksilver. Electricity be my metaphor! (with apologies and thanks to Dylan Thomas).

McLuhan did not use terms like "general systems theory" and "team learning," but he embraced these notions using terms like "cybernation" and "feedback." "Feedback is the end of the lineality that came into the Western world with the alphabet and the continuous forms of Euclidean space" (McLuhan 1964, 354). He also stressed interrelation and the need to reverse the fragmentary nature of industrial-age education. "Education designed to cope with the products of servile toil and mechanical production are no longer adequate. Our education has long ago acquired the fragmentary and piece-meal character of mechanism. It is now under increasing pressure to acquire the depth and interrelation that are indispensable in the all-at-once world of electric organization" (McLuhan 1964, 357).

Senge stresses the importance of research within a learning organization. All solutions are "always subject to improvement, never final . . . When people in an organization come collectively to recognize that nobody has the answers, it liberates the organization in a remarkable way" (Senge 1993, 282). He saw the role of the manager in a learning organization "as researcher and designer" (Senge 1993, 288). McLuhan also recognized the importance of research and the participation of the staff of an organization in that process through his interpretation of the Hawthorne effect. Industrial psychologists observed at the Hawthorne plant of General Electric that no matter how workers' conditions were changed their productivity increased. McLuhan attributed this to the fact that their participation in the experiment motivated them and made them work harder.

NOMADIC GATHERERS OF KNOWLEDGE

McLuhan certainly demonstrated an uncanny ability to describe the landscape of cyberspace, a term I am certain he never heard. What is remarkable about his achievement is that in 1964 when he wrote *Understanding Media*, workplace computing was in its infancy and restricted to mainframe computers in which jobs were run on a batch basis using punched-hole cards and card readers. His musings on computers filled fourteen pages of the last chapter of his book and two pages in the introduction, yet he seems to have covered the gamut of issues facing computing and learning in today's workplace. He even seems to anticipate, albeit obliquely, the phenomenon of surfing the Internet. He suggested that with electricity and automation, human behavior changed. "Men are suddenly nomadic gatherers of knowledge, nomadic as never before — but also involved in the total social process as never before; since with electricity we extend our central nervous system globally, instantly interrelating every human experience" (McLuhan 1964, 358).

IN SEARCH OF LEARNING ORGANIZATIONS

McLuhan, Toffler, Hammer and Champy, Tapscott and Caston, and Senge agree on the critical role that knowledge now plays in the creation of wealth. Their observation is both trivial and profound. It is trivial in the sense that it is a cliché which no one would deny, yet profound because there has been no significant response by the education community and only a minor one by the business community to this new reality transforming our lives. Despite the dynamic changes in the nature of our economy, the training and preparation of personnel are based on the industrial model. The world which our present forms of organization and our education system address is rapidly disappearing, yet the way our school system relates to the world of work has not changed significantly since its founding at the beginning of the Industrial Revolution. The classroom has traditionally been isolated from the basic realities of work and the economy. The stereotype of teachers pedantically dispensing book learning unrelated to the needs, concerns, or interests of their pupils is, regrettably, far too accurate.

Nevertheless, the school system demonstrates some measure of success in that it supplies technically trained individuals. Unfortunately, these successes are limited. Today's educational goals serve yesterday's needs. They are built into the system because education was organized to supply the work force for a hierarchical society. Its function was to stream the general population into an elite of highly specialized experts and technocrats, the managing class, and a general population with enough literacy and basic skills to consume industrial products and provide a disciplined labor force. The hierarchical structures of industrial organization are rapidly disappearing, however, and therefore the high-end products of the school system — the middle managers — are no longer needed.

This is slowly being recognized in the private sector and is resulting in a reorganization of firms along the lines outlined earlier and a reengineering of their business processes. There is a flattening of the hierarchical management structures and a movement away from command-and-control management. The public sector has been much less responsive to this form of reorganization than the private sector, but the idea has been gaining general acceptance. Least responsive to these trends has been our education system, perhaps the most reactive and hierarchically structured institution in our society.

We can no longer afford to entertain an educated technocratic elite who take the responsibility of operating and administering economic, political, and social institutions for the less educated lower echelons of society. The fragmentation of knowledge among different social classes cannot succeed in the computer age as it did during the industrial era when tasks were also fragmented and only managers needed to be fully informed.

Today's economy demands greater integration of job function with knowledge and information. "The 'leap' to a higher level of diversity, speed, and complexity requires a corresponding leap to higher, more sophisticated forms of integration. In turn, this demands radically higher levels of knowledge processing" (Toffler 1990, 81). There are two implications to this observation regarding the training of postindustrial workers. First, the present practice of dividing the school curriculum into specialized packages of information called subject disciplines will no longer serve today's need for a systemic approach to the use of knowledge. According to McLuhan, "Automation is information and it not only ends jobs in the world of work, it ends subjects in the world of learning" (McLuhan 1964, 346). Secondly, all members of society, whether they are production workers, service sector workers, or consumers, will require both a knowledge base and a set of information-processing skills to function at even the most rudimentary level. Clearly, our school system must be capable of preparing individuals for the new demands of postindustrial society.

Toffler's analysis of the economy and society as developed in *Powershift* can be applied to education and provides an explanation for the present failure of our school system: "The disintegrative or analytic approach, when transferred to economics, led us to think of production as a series of disconnected steps. Raising capital, acquiring raw materials, recruiting workers, deploying technology, advertising, selling, and distributing the product were all seen as either sequential or as isolated from one another" (Toffler 1990, 81). Toffler's description of industrial production can be paraphrased to diagnose with equal acuity the problems with today's education system: "The disintegrative or analytic approach, when transferred to schools, has led us to think of education as a series of disconnected steps. Reading, writing, mathematics, science, social studies, computer literacy, commerce, engineering, medicine are all seen as either sequential or as isolated from one another." The origins of reading, writing, math, and science are all connected and hence these disciplines should be studied in an integrated manner. An integrated curriculum not only better prepares people for the reality of today's economic and commercial climate, it also makes learning easier.

The lesson of the integration of learning and doing does not apply solely to our schools. It applies with equal validity in the workplace, in which job function and training must also be intermeshed. Industry has done little to reform its educational practices. It was forced to introduce training programs for staff and in some cases operates in-house training departments. These training departments have been formed more often to save on the cost of training than out of a real commitment to employee training and development. Most in-house training departments operate at a level below that of outside organizations whose sole business is education. In the field of computer training, for example, an in-house training group will confine its expertise to the software and hardware platform that the organization is using. They may read about other computer products but they do not immerse themselves in their operation and therefore do not have the same breadth of understanding of the current computing environment as trainers in an educational company working with a large number of software applications and platforms.

Industry still operates on the antiquated premise that the basic education of employees has been provided by the school system and their only responsibility is to upgrade the skills of employees from time to time. In too many instances, in-depth education of personnel is viewed as a waste of money and resources. The cost of training is often looked at as an additional expense, like taxes, that must be minimized no matter what the consequences — and the consequences can be significant. Since wealth production and knowledge are synonymous, the first price to be paid is the inefficiency in wealth production due to personnel who do not have the basic skill sets needed to do their jobs. The analogy in the industrial age would be to consider the hardware or machinery necessary to operate a factory as an ancillary cost of doing business that needed to be reduced as much as possible or avoided altogether.

Industrialists realized that the advantage their factories would have over their competitors would be the efficiency of their equipment and machines. Today's chief advantage is not just the equipment, which is still critical, but the knowledge base of one's employees at both the staff and management levels. The early industrialists understood the importance of their machinery: its results were so tangible. The unit cost of production using machines was dramatically lower than hand production, making the need to invest in equipment obvious. The only exercise was to calculate the time needed to recoup the capital investment. In the early days of industrialization, marketing and competitive analyses were not even required because there was little or no competition and an undersupply of consumer goods, so it was a seller's market. Undercutting the price of handcrafted goods was a no-brainer.

Obviously, recouping the investment made in employee education and development is not as easy to gauge as hardware investments were in the industrial age. Increased productivity due to training may be marginal at first, so it is difficult to calculate the payback time. Given the low cost of training, however, relative to the cost of hardware, operating costs, and salaries, a modest incremental increase in productivity will result in a payback of only two to six months. Let us estimate the payback time for a typical training scenario, assuming that the cost of training per day is twice the employee's daily salary and the training session lasts two days. The cost is therefore six paydays, four for the actual cost of the training and two for the lost productivity during the two days of training. A modest increase in productivity of 10 percent due to the training will result in a payback period of only sixty working days or roughly three months. No manager should ever think twice about providing each employee with a minimum of two days of training every six months.

The other benefit accrued from training is the general intellectual and cognitive development of an organization's employees. Not only will they be able to do their jobs more effectively, they will also be better able to develop new ideas to make their firms more competitive and/or more productive. A knowledgeable employee is a more effective employee and also a happier employee. When workers' skill sets are being continually updated, they feel more confident about their work. They have pride in what they are doing. They also are more motivated because they feel that their companies care about their well-being and personal development. Employee training and education not only make for a more knowledgeable worker but also a more motivated one. Even if employees were thoroughly trained for their jobs, excuses should be invented to train them just to increase their sense of participation and harness the Hawthorne effect (McLuhan 1964, ix).

Finally, the bottom line for training employees is the need to remain competitive. Progressive firms are benefiting from the enormous dividends that employee education pays. It will not be too long before the rest of the business world catches on to the obvious: that in an information-based economy, knowledge is the prime ingredient in the production of wealth. The number-one asset of a company, therefore, is not capital, not equipment, but knowledge and a well-trained staff — intellectual capital, if you will, and that intellectual capital must be spread uniformly throughout the organization. This is in striking contrast to the industrial age, where the knowledge sat at the top of the pyramid and was carefully guarded and hoarded by the owners and managers, and a program of de-skilling employees and segmenting work was considered the best way to increase efficiency.

Toffler's description of the new production paradigm can be easily adapted and used as a prescription for remedying the ills of today's school-based and workplace educational practices. We need only substitute the word "education" for "production" in the following quote of Toffler's, from *Powershift*: "The new model of production [education] that springs from the super-symbolic economy is dramatically different. Based on a systemic or integrative view, it sees production [education] as increasingly simultaneous and synthesized. The parts of the process are not the whole, and they cannot be isolated from one another" (Toffler 1990, 81).

The skills learned in one discipline must feed the study of another. Topics must be studied holistically where the literary, mathematical, scientific, social, economic, and environmental aspects of the topic are treated simultaneously. "Real-time" or experiential involvement in a topic must supplement book-based learning. "Connectivity rather than disconnectedness, integration rather than disintegration, real-time simultaneity rather than sequential stages — these are the assumptions that underlie the new production paradigm" (Toffler 1990, 82), and by extension the new education paradigm.

Toffler describes the impact that the new mode of production is having on business and the distribution of knowledge: "Just as we are now restructuring companies and whole economies, we are totally re-organizing the production and distribution of knowledge and the symbols used to communicate it" (Toffler 1990, 85). Unfortunately, a similar restructuring of our education system has not yet taken place and this is why our schools and industrial training continue to fail us.

When I refer to the failure of our education system, I am not referring merely to the system of publicly and privately funded schools to serve the primary, secondary, and postsecondary educational needs of the community. My remarks also include the failure of the industrial training programs in both the public and private sectors, as well as the lack of integration between the training provided by the formal education system and the needs of the workplace. And even if we succeed in reforming the system, the rate of change of technology is still too rapid to rely on schools to meet all our educational needs.

To succeed in its mission, today's business organization must give careful thought to its education strategy. It must strive to become a "learning organization, where people continually expand their capacity to create the results they truly desire, where new and expansive patterns of thinking are nurtured, where collective aspiration is set free, and where people are continually learning how to learn together" (Senge 1993, 3). Unfortunately, most of our educational institutions, even those of higher learning, are not "learning organizations" by Senge's definition. More about that in the next chapter.

7

THE STUDENT IS THE CURRICULUM AND THE PROCESS IS THE CONTENT: Computers and the New Curriculum

COMPUTERS: COGNITION, COMMUNICATION, AND EDUCATION

The ostensible objective of an education system is the transmission of a body of knowledge, but equally important is the transmission of the methodologies and techniques for organizing information. In fact, as noted earlier, the heart of an education program is not the information with which one becomes acquainted but the communication, information-processing, problem-solving, and cognitive skills that are developed or acquired through understanding a body of knowledge. The histories of education, cognition, information processing, communication, and language are therefore intimately connected and must be studied together. "Language and human intelligence can be thought of as two aspects of the same evolutionary development" (Hickerson 1980, 13).

In fact, information-processing skills are basically equivalent to cognitive skills as the following definition by a cognitive scientist suggests: "Cognitive psychology attempts to understand the nature of human intelligence and how people think . . . Cognitive psychology is dominated by the information-processing approach, which analyzes cognitive processes into a sequence of ordered stages" (J. R. Anderson, 1980, 3).

In order to study the role of computers in education, it will be necessary to study the three domains of communication, education, and cognition (or information processing). We must also examine how these three domains relate to one another and how they are affected by technology. The boundaries between the three domains have begun to blur partly because of the integrative power of computers. Perhaps, then, cognition, communication, and education are not the three separate categories that the analytical objectivist approach of Western science and social science, with their bias towards classification, would have us believe. Perhaps cognition, communication, and education are simply different projections or facets of the same phenomenon — human thought.

It is difficult, for example, to imagine one of these activities being carried out without the involvement of the other two. Indeed, none of them can be pursued independent of the other two. Cognitive activities require both the communication of information and the skills acquired through education. The linguists Edward Sapir and Benjamin Lee Whorf argue convincingly that human thought or cognition sprang from humankind's unique capacity for speech or language. As children begin to develop mentally, many of their cognitive processes are learned within the context of their social and information environment. Hence, cognition is a product of both communication and education. Communication, on the other hand, requires both the development of cognitive skills and participation in an educational process. Language, too, is learned from the child's interactions with his or her parents and family. The development and deployment of a child's communication skills, however, also depends on his or her cognitive skills. Communication is, therefore, also a product of the domains of education and cognition. Finally, education is not possible without both the communication of ideas and the deployment of cognitive skills. My argument that each domain depends on the other two comes full circle with this last observation.

The relationship between cognition, communication, and education is impossible to formulate in a reductionist manner because the interactions between these three domains are so strong and interwoven. I believe these processes arise together, each creating the conditions for the other's development.

Any two of the domains in this model can function as inputs into the mind to create or generate the third. It is impossible to have one without the other two, but none of the three domains is primary. As one considers different intellectual activities, the domains which function as the inputs in our model and the domain which will function as the output shift. If the activity is learning, then the inputs are cognition and communication and the output is education. If the activity is sharing ideas, then the inputs are cognition and education and the output is communication. If the activity is thinking, then the inputs are education and communication (or language) and the output is cognition.

The link between cognition, communication, and education is related to the fact that language has both a communication and an informatics aspect. The Russian psychologist L.S. Vygotsky formulated a similar relationship between language and thought, which applies with equal validity to the relationship between communication, cognition, and education.

> The primary function of speech is communication, social intercourse. When language was studied through analysis into elements, this function, too, was dissociated from the intellectual function of speech. The two were treated as though they were separate, if parallel, functions, without

> attention to their structural and developmental interrelation. Yet word meaning is a unit of both these functions of speech . . . The conception of word meaning as a unit of both generalizing thought and social interchange is of incalculable value for the study of thought and language. (Vygotsky 1962, 6–7)

Vygotsky is criticizing the separation of the communication or social-intercourse aspect of language from the intellectual or cognitive function, and is pointing out how this limits our understanding of the relation of thought and language (or, as I have formulated the problem, the relation of communication, cognition, and education).

Technology and the Transformation of Education

The expansion of the notion that language is simply a medium for communication to include its informatic abilities to store, retrieve, organize, and process information is essential for education and curriculum planning. Contemporary education does not limit itself to the teaching of the 3 Rs. It attempts to develop the student's reasoning and problem-solving capabilities; thus, the information-processing aspects of language are critical for a well-rounded education. In fact, education can be defined as the learning of the use of the five languages of speech, writing, mathematics, science, and computing. The emphasis in this approach is not learning facts but using these languages to study one's world and to organize and communicate information about that world. This is why computing, with its emphasis on the information-processing aspects of language, has the potential to revitalize and revolutionize education.

Our model of language as communication and informatics can be extended to describe the education system. Schools have both a communication and an informatics side. Teachers and students communicate with one another, while the significant part of the curriculum concerns itself with teaching various forms of information processing.

Historically, each time a new medium of communication or method of information processing has been introduced, education has been directly affected and, in many instances, radically transformed. For example, the idea of formal education housed in an institution devoted to learning — a school — is a direct result of the invention of writing and the desire to train children to be literate. The advent of writing and numerical notations signaled the end of an education system based on the oral tradition and the apprenticeship mode of vocational training. Education was a natural part of everyday life in preliterate societies and was integrated with the normal activities of survival such as homemaking, toolmaking, hunting, and food gathering. The new system of education that emerged with writing, however, was based on formal schooling and the instruction of a specialist, the teacher, whose sole mandate was to provide formal instruction. The education of children was separated from the other activities of society and housed in a new institution, the school.

As discussed, although the microcomputer has had hardly more than fifteen years to influence education, its impact has been significant. This technology has enjoyed great success in helping learning-disabled children, particularly those who are dyslexic, overcome their handicaps. Its role in the general classroom is extremely promising and still emerging. It is already clear, however, that the microcomputer's capacities to motivate students, encourage peer teaching, individualize instruction, and increase user interactivity hold the potential to revitalize education at both the grass-roots level of the individual student, as well as at the curricular level throughout the whole system.

Formulating a Curriculum of Study for the Information Age

There are two areas where change will be required in the education system: the actual content of the curriculum of study and the way in which that curriculum is presented or the way in which the curriculum will be addressed by students. The concept of the delivery or the presentation of a curriculum denotes an attitude and style of education characteristic of the industrial age in which schools operated more or less like factories. This discussion will examine the delivery or presentation of curriculum in the spirit in which the student proactively addresses a curriculum of study rather than passively consumes it as was the case with the mass education system of the industrial era. The typical schools of the industrial era have been likened to sausage factories where students were stuffed with knowledge and information. Each student was stuffed with exactly the same material so as to produce a uniform product. Today we recognize the shortcomings of such an approach and try to individualize education as much as possible so that each student may realize his or her own unique potential.

Educational psychologists have discovered that children (and adults) learn better when they have an opportunity to choose what they study and when they study it. According to the developmental psychology of Jean Piaget, "Learning is an active process whereby the child constructs understanding of events in his or her environment by a process of assimilation and accommodation. The child assimilates new experiences into existing cognitive structures for understanding the world. These structures, in turn, accommodate themselves to the new experience and thereby slowly mature into the complex reasoning abilities that characterize formal adult thought" (Kurland and Kurland 1987, 322).

The child must exercise some element of control over what is studied and when. This is not to advocate total freedom of choice for the student. If this were the case, there would be no need schools. Libraries and curricular resource centers and occasional visit of a tutor would satisfy all of the student's educational needs. There are shortcomings with the sixties style of education or the human potential model, in which students are free to realize their potential by studying only what interests them. While the student's participation in the selection of the topics or subjects of study is essential for the success of any educational endeavor, education requires structure and the guidance and devotion of a caring adult. A balance must be struck between students pursuing their own educational interests and goals and educators providing guidance and structure. There is no inherent reason that students cannot pursue the goal of self-actualization while preparing themselves for living and working in society. The style of education and the specific curricular goals advocated here strike a balance between the needs of the individual and those of society.

CURRICULUM BY TRADITION

Tradition plays a dominant role in the organization of learning in our education system. There is no universally agreed-upon theoretical basis for the present curriculum of our primary and secondary schools. To some degree, today's curriculum is nothing more than an historical accident with little or no rationale for its organization. Moreover, it is not clear that contemporary teaching practices introduce skills at the most appropriate time in the development of pupils.

There are isolated attempts to incorporate some of the insights of educational psychology into curriculum design. The insights of cognitive science tend to dominate, but elements of behaviorism still linger. One cannot blame the educators with the responsibility of determining curriculum for not making greater use of the insights of educational psychology. The field is in flux, with the answers to many vital questions violently debated by diverse schools of thought. In addition, educational psychology is subject to fads and bandwagon effects, which makes educational planning difficult, if not outright impossible. The lack of consensus among the experts generates just enough confusion and uncertainty among educational planners, curriculum designers, and school board trustees to neutralize the contribution that educational psychology could make.

A NEW APPROACH TO CURRICULUM PLANNING

Rather than attempting to plan curriculum on the basis of the insights of educational psychology with all of its uncertainty, I propose a much simpler approach based on an understanding of the nature of language, communications, and informatics. Rather than trying to fathom the mysteries of the cognitive domain, I suggest that we allow the historical development of language, communications, and informatics to guide our philosophy of education. The natural evolution of language provides a model of how and when to introduce concepts and techniques in communications and informatics. This approach is pragmatic and phenomenological rather than theoretical.

The basic assumption of this philosophy of education is that the historical course in which cognitive processes develop within the human psyche can be successfully repeated in the education of the individual. The basic idea can be expressed in terms of the concept used in the field of biological evolution, that "ontogeny recapitulates philogeny."

In essence, I propose that formal education focus on the integrated development of the communication and informatics skills involved in the use of the five verbal languages of speech, writing, mathematics, science, and computing. Conceptual tools or cognitive skills that arise from the use of the earlier forms of language will prove useful in mastering the later modes of language. Thus, the tools of analysis and classification acquired through learning writing and mathematics can be used in the study of science and computing. The primary focus of education becomes the development of the generic cognitive skills associated with the five modes of language, whereas the mastery of any particular body of knowledge becomes a secondary goal. Examples of generic cognitive skills include analysis, abstraction, classification, matching, measuring, coding, decoding, metaphor creation, model making, synthesis, and so on. By adopting an integrated approach to education, students will not only acquire the communication and information-processing skills associated with the five languages but they will also develop the problem-solving skills and creativity necessary to apply these skills to new situations.

This model of education is based on the notion that all modes of language must be mastered and incorporated into an individual's repertoire of skills. Some would argue that, with the exception of those who want to become computer scientists or office workers, learning to use computers is a frill for most students. They would argue that the use of computers is a luxury that most schools and jurisdictions cannot afford. I believe that the opposite is true and that computing is just as essential to a complete education today as learning how to read and write was after the advent of literacy.

Including Computer Technology in Today's Curriculum

At each stage in the evolutionary chain of language, stretching from the spoken word through writing, mathematics, and science to computing, the information-processing capacity of human language has increased. Each new development of the language was motivated by the need for a higher level of informatics to deal with or combat the information overload that arose from the previous mode of language. Therefore, it is inevitable that as each new form of language emerges, it is adopted by the information elite of society. It is also inevitable that not only will the new form of language help alleviate the information overload of the previous mode, it will also create new applications of language and thus eventually give rise to new forms of information overload. Computers, which were developed by science and business to cope with the enormous amount of data generated by the success of industrial society, have given rise to a new information explosion in the postindustrial era. Once a luxury or a convenience, computers are now a necessity. They are here to stay and their use will only increase with time.

Once a new form of language emerges, one can never permanently revert to the information environment that existed before its advent. There are examples in history where a culture falls into a period of temporary intellectual regression and reduces its use of one of the more advanced forms of language that had been previously adopted by the culture. The use of writing in Europe declined almost totally with the fall of the Roman Empire. It was only in the monasteries that book learning and the art of writing were preserved in Europe. Periods of cultural decline, however, are generally followed by an intellectual restoration or renaissance in which the earlier developments are recovered and return even more vigorously than before.

Each time a new form of language was invented, it quickly established itself as the dominant mode of communication and informatics in certain domains of intellectual activity. The dominance of the new mode of language does not obsolesce the earlier forms of language, however. First, the new mode of language automatically incorporates or subsumes all of the earlier modes of language, hence preserving them. Secondly, the new mode cannot successfully compete with the older forms of language in every possible application. The complexity that the new mode of language must possess in order to provide greater informatics power makes it awkward to use in certain circumstances. The older modes of language are therefore absolutely essential to human intercourse and continue to evolve but do so under the influence of the later forms of language. For example, people did not stop talking to one another after the advent of writing. But the nature of the spoken word did change as a result of the advent of writing. A comparison of the vocabulary of Homer, representing the oral tradition, with that of the postliterate period reveals that more abstract terms crept into the ancient Greek language with the advent of writing. Although today computers pervade almost all aspects of communication and information processing, we do not use a computer to write out a shopping list or to perform a simple mathematical calculation like balancing our checkbooks. Computers, however, have changed the way we prepare written documents and even the way we communicate informally with colleagues by using e-mail.

The Five Languages: A Core Curriculum

Because each mode of language has its own set of unique domains of applicability and areas of practicality, a well-educated person needs to be adept in all five. A lack of knowledge of any of the five modes restricts an individual's sphere of activity. Given that the purpose of an education is to allow individuals to realize their full potential, students should be exposed to speech, writing, mathematics, science, and computing. The degree to which any given student will want to study each of the five modes will depend on his or her needs, interests, and aptitudes. Mastery of the rudiments of all five is certainly a prerequisite for serious study in most areas of learning.

Learning to speak is not a part of formal schooling, as virtually all children learn this skill before they reach school age from their parents or guardians. Refining the skills of oral language or learning the art of rhetoric, however, is an appropriate part of the school curriculum. Therefore, the main elements of a formal education are:

1. Rhetoric — the art of public speaking.
2. Literacy — the ability to read and write.
3. Numeracy — the ability to use abstract numbers and perform simple mathematical operations.
4. Science literacy — the ability to use scientific principles to organize knowledge and interpret information.

5. Computer literacy — the ability to use computers to communicate, store, access, organize, and process information.

These definitions will be refined and elaborated upon in greater detail later. For the moment, they provide the framework for formulating a well-rounded and comprehensive program or curriculum of formal education.

An education does not consist merely of data handling and information-processing skills. While developing the analytic skills of its students must be a primary concern of an education system, there is still the question of how students will put these skills to use and what things they will value in life. A school system cannot be content just to create technocrats who can operate the machinery of industry and society. The core curriculum I have proposed leaves out a number of activities and studies that should be included in a total educational environment; for example, health and physical education, life and coping skills, values education, social skills, and exposure to and participation in the nonverbal languages and arts such as music, dance, painting, sculpture, photography, video, handicrafts, and so on. I fully recognize that these activities are essential to a comprehensive education and hence must complement any core curriculum which focuses on the five verbal languages. I will not deal with the nonverbal languages and the other topics in this study, however, as my focus is language.

Still, it must be noted that these areas in the student's development are extremely important and should not be considered as add-ons or mere cultural enrichments; rather they must be integrated into the core curriculum that I am proposing. Educational reform must go beyond reconverting our schools from factories which produced skilled workers for the industrial age to one that produces them for the information age. We should strive to turn out well-rounded students who are both technically skilled and socially and culturally aware.

PROCESS OVER CONTENT

The structural interrelatedness of the five modes of languages identified in Chapter 3 suggests a design for the core curriculum for children. The actual content of the curriculum, that is, the domains of knowledge or bodies of information that are studied, are not as essential as learning the techniques of informatics and communication associated with the five languages. Reading can be learned from schoolbooks or just as easily by reading the instructions to a board game or a computer program. Writing can be learned through personal expression, storytelling, or reporting on a project. The principles of science can be learned by studying animal behavior, plant morphology, geological formations, or the phenomena of electricity and magnetism. Computing skills can be acquired through word processing, database management, or numerical analysis.

The choice of the topics to be studied using the five modes of language should be based on the interests and needs of the pupils and the community in which they live. Obvious choices for topics are the most conspicuous features of the students' environment and the most dominant elements of their culture, such as national and local history and geography. It would also include local industries such as agriculture in rural communities, the ocean in maritime communities, forests in a lumber community, manufacturing and commerce in an urban setting, and so on.

The order in which the various modes of language are introduced should be the product of careful study and analysis within the domain of cognitive science. Until cognitive science can provide more direction, certain common-sense rules of thumb should be followed. Each new mode of language should be introduced at an appropriate time in the education and development of a youngster. The historical development of language can guide curriculum developers. This does not mean, however, that the pupil must acquire a sophisticated grasp of one mode of language before being introduced to another. In fact, the use of one mode can help in mastering another. For example, interest in science or computing can be used to teach reading and writing skills; the need to solve a mathematical problem can become the impetus to learn more about computing.

Integrating the use of the five languages will, in fact, help ensure the success of the curricular approach I propose. The computer provides a natural medium for the integration of the five languages because it incorporates elements of all the others. It is also a natural medium for integration because of the ease with which it can store and manipulate information.

Using all five modes of language to study a particular topic is a natural way to achieve integration. It also corresponds to the way problems are addressed outside of the classroom in the "real world." Reading, writing, spelling, mathematics, science, and computing should not be regarded as subjects to be taught; they should be treated as tools to be mastered and used to learn about things that interest the students. These tools become the windows through which students can study the world in which they live and the media through which they communicate that knowledge to others.

According to educator John Olson, this approach to the tool use of computers in the general curriculum also has the effect of integrating two important streams of the educational process: vocational training and general education. "Computers can contribute both to educational and social goals. The study of them can introduce students to the 'world' of computing, increase computer awareness and develop computer literacy — all of which are aimed at coping in the working world. Computers can aid the teaching of existing subjects, or they can introduce students to whole new ways of learning. Thus we have two important dimensions to keep in mind; vocational–educational and school subjects teaching aid" (Olson 1988, 2).

I prefer to think of these two dimensions in terms of the vocational training and cognitive development of students. There are some aspects of learning about computers which are primarily vocational but which also will contribute to the student's cognitive development. These include learning how to program, design, or repair a computer. These specific toolmaking skills, as opposed to tool-using skills, would prepare those students who wish to make a career out of computers.

The tool-use skills of the computer provide the student with general vocational preparation as well as an excellent tool for general learning and cognitive development. Given that so many more information transactions will take place through the medium of the computer, the dichotomy between the vocational and educational uses of word processing, spreadsheeting, or databasing is moot. The skills of computing are no more vocational or educational than reading and writing. They are both. Many aspects of education and many areas of employment are denied to anyone who does not learn to read and write. The same is true for computing and will be even more so in the future. The richness of education and the possibilities for work will be limited for those who do not learn computing. Today, many workers are being displaced because they cannot use a computer. Thus the integration of computing, particularly the use of computers in the tool mode, will serve both an educational and a vocational function for today's students who will live their lives entirely in the information age.

The real lesson to be learned is the art of uncovering or discovering new knowledge, the art of organizing that knowledge and applying it to practical situations in life and, finally, the art of creating new knowledge or conducting original research. What is being proposed here is not really that different from present-day views on the curriculum; it is more a change of emphasis and a reversal of the roles played by content and technique. This approach also attempts to provide the student with power over or control of the technology rather than the reverse situation (Labbett 1988, 91).

If we are to help students become more experienced in using information technology to pursue their own educational goals, then the curriculum aim hinges on the successful design, management, and assessment of a classroom curriculum, where the teacher and his/her pupils can jointly develop their understanding of the world (Labbett 1988, 90).

My approach to curriculum planning can be expressed in terms of two of McLuhan's aphorisms: "The medium is the message" and "The user is the content." The "media" of education are the five languages, and the "content" is the various topics to be studied, which are grist for the mill of honing the skills of using the five languages. Rather than learning the three Rs in order to study other subjects, other subjects are studied in order to master the three Rs, science, and computing. As "the user is the content," the subjects selected to be studied should be those that interest the student.

TUTOR, TOOL, AND TUTEE MODES OF COMPUTING

In a cogent formulation of the role of computers in education, educator R.P. Taylor (Taylor 1980) identified the tutor, tool, and tutee modes of computer usage. In the tutor mode, the computer is used as an automatic teacher, instructing the user by delivering information, requesting responses, and matching the student's response to what might be considered the correct answer. This mode is sometimes referred to as CAI, computer-assisted instruction, or CAL, computer-assisted learning. Simulations in which natural phenomena or social interactions are modeled represent another example of the tutor mode.

In the tool mode, the operator uses the computer in order to achieve some information-processing or communication task. Examples include word processing, spreadsheets, database management, or graphics. In the third mode, the tutee mode, the user is teaching the computer or instructing it what to do. Programming the computer is an example of the tutee mode. The tutor, tool, and tutee modes of computer applications in education each have merit. Educators must address the question, given the limited resources available, what are the most powerful and practical uses of computers? I believe that the tool mode is the most effective way to use computers in education because it allows the user to master the information-processing and communication skills associated with the language of computing.

Using a computer in the tutee mode contributes to the cognitive development of the user. Programming a computer is more difficult than using a computer in the tool mode and is also more specialized. It is the kind of skill a computer scientist will need to master but it is less useful to the ordinary end user than learning a tool application. Some exposure to programming or related activities like LOGO is advisable, however, so that a student has a better understanding of computers and how they work. Using a computer in the tutor mode does not contribute to the cognitive development of the user in the way the tool and tutee modes do. In addition to being of limited value, it is not clear that CAI has any advantages over traditional methods of drill and practice. "Not all studies of the effectiveness of CAI have shown consistent positive results . . . The researchers found no strong evidence that PLATO [a CAI program] either helped or hindered students . . . The studies have also consistently pointed out that the cost-effectiveness of CAI programs — one of CAI's major selling points — has clearly not been demonstrated" (Hickerson 1980, 337, 343).

CAI reduces the user to a passive consumer of information that is being automatically presented to him or her by the computer. There are benefits to be derived from learning the contents of the material being presented in the tutorial mode, but the student does not derive any additional benefits from the process of using the computer this way. It is similar to television use, which is informative but does not develop the skills of its viewers. There is also the concern of the overexposure of youngsters to video images (Sloan 1985, 7–8). The use of the computer in the tutor mode must have demonstrable educational benefits to justify its use. It would be hard to support the use of CAI where it was the sole or primary use of the computer.

The tutor mode also includes simulations in which a software program models a natural phenomenon, a complex piece of technology, a social interaction, or a social institution. By exploring the model, the user begins to learn about the phenomenon that has been modeled or simulated. Simulations can have a negative effect if they are used to replace experiences that can be had in the real world. On the other hand, they can be quite useful if they can show students how a system they could not otherwise access operates, such as a nuclear reactor.

If a simulation is to be used as an effective educational tool, it must be integrated into the student's overall educational program. Students and their teachers should discuss how well the simulation models reality. Field or laboratory comparisons with the processes being modeled should be attempted wherever possible. The skills that are being developed through the software program and the topic of the simulation should be studied using other educational media such as texts and discussions. In short, simulations should be chosen for use if they reinforce other classroom activities and can be integrated into the curriculum effectively.

Of the three modes of computer usage in education, the one that I most strongly advocate is the tool mode, because it promotes cognitive development, is easy to learn, and has many practical applications. The use of computers to organize information derived from the direct observation of real phenomena provides the healthy mix of using a powerful analytic tool and at the same time dealing directly with nature. I propose that students use computers as tools, just as they are used in the real world of work, whether in the field of art, science, administration, commerce, or engineering. The computer can be used as a tool to capture the fleeting images of the direct observations of the real world so that the data may be examined, reexamined, and interiorized.

In summary, then, all applications of computers have merit. The challenge to educators is to choose the appropriate mix of uses which best meets the educational needs of the students for which the educational program is being designed. It is recommended that the tool mode be given the greatest priority because the tool use of computers (1) promotes cognitive development; (2) reinforces information-processing and communication skills; and (3) provides an open-ended environment for imagination, exploration, and discovery. According to one study,

> word processing . . . is already a sophisticated tool for the production of reports, essays, poems, books, etc . . . It offers an open-ended facility for the imagination. It aids revision through a variety of techniques and presents the finished product in a polished form . . . The product can then be stored, say on a database, or transmitted to others on a computer network. Similarly spread sheets offer facilities for computational analysis . . . Other open-ended tools offer the facility to produce diagrams and graphs . . . The computer offers, therefore, the chance to make real products for an audience, an audience not just limited to classmates or teachers. (Kurland and Kurland 1987, 16–17)

The tool mode of computers will be discussed in much greater detail in the Appendix; one section there is devoted to the role of word processing and another to the role of spreadsheets and databases, with special emphasis on their application in the natural and social sciences.

Computers: A New Source of Alternatives in Education

One of the solid achievements of the period of intense social change during the 1960s was the emergence of alternative schools. The uniformity of traditional schools, which were organized in conformity with the values of the industrial age, proved to be oppressive for many children and adults. Some children who were labeled learning-disabled within the context of traditional schools thrived in alternative educational environments. Alternative educational formats which were initially resisted by traditionalists are now part and parcel of the services offered by most school boards to serve the total educational needs of their community.

As already noted, microcomputers will be a source of new alternatives in education. They provide an alternative to traditional textbooks as deliverers of information which can be easily integrated. The ease with which information can flow between computers, or between software programs on the same computer, or between different files within the same program, facilitates the integration of a variety of subjects or topics.

The computer also provides children with an alternative channel for inputting and outputting textual information. The work of special education teachers with learning-disabled children and computers has proven beyond a doubt that computers offer many children who have difficulties dealing with print the opportunity to develop a functional level of literacy. Many children who have been labeled learning-disabled are not disabled if an appropriate mix of media is used to facilitate the development of their literacy. The microcomputer seems to be an important component of that mix.

Other teachers working with populations of "normal" students systematically report that computer usage leads to a greater competency with literacy skills compared with traditional methods. A similar increase in mathematical skills has also been reported. Many cases can be cited where the use of computers has not led to any significant educational or cognitive gains, but this is most often due to the misuse of the technology by the teacher. There is little doubt that computers offer an important alternative to the traditional approaches of teaching the 3 Rs.

Although it does not work for everyone, the computer is a very powerful learning tool for many children and adults. There are some people for whom the traditional methods of textbooks, pencil, and paper may be more suitable. And for still others, neither the traditional approach nor the use of computers is the answer. For these individuals, the educational techniques of the oral tradition may be the most effective. And for still others, the medium which taps their potential may be one of the nonverbal media of expression and communication such as music, dance, painting, or theater.

There is more than one way to become literate. I know of a student labeled learning-disabled by the system and made to feel like a failure who found his own path to literacy through music. This individual, who dropped out of grade 9, first became a rock-and-roll drummer and later a symphonic percussionist. He eventually played with two of North America's leading youth orchestras, began a career as a composer, and graduated from a prestigious music school. He not only achieved a university level of literacy, but earned an A average because of his passion for his musical art form. In addition to his achievements in musical performance and symphonic and film-score composition, he went on to master computer programming in order to compose and record his music, became an avid reader of philosophy to better understand his nonverbal art form, and began writing poetry, a form of verbal expression closely related to his interest in music. If he had remained in the education system and not dropped out when he did, he might have been totally destroyed as a musician and as a person. The important lesson to be learned from this is that an interest in one medium of expression can lead to proficiency in a number of other media.

Not only are microcomputers in their present form an alternative medium of expression and education, they also hold the potential for becoming the source of many additional new modes of communication and education. By integrating graphics and/or music with textual materials through multimedia and hypermedia packages, microcomputers can integrate the educational experiences associated with verbal and nonverbal languages. Perhaps computers can be used to explore the relationship of music to mathematics and physics or the relationship of the fine arts to geometry and optics.

As we have seen, another important educational alternative that microcomputers promote is individualized learning. Clearly, the uniformity of the mass education system that served the needs of the industrial age no longer serves the needs of the information age. Individualized instruction is required, but we still want to provide an education for all sectors of our society, as well as make education a lifelong activity. The only way individualized education can still be provided in the context of the increased demand on the education system due to increasing population and the desire for lifelong education is to make wholesale use of computers and integrate them with classroom instruction so that the important social exchanges of peer interaction and teaching can be maintained.

Defining Computer Literacy

We will require a definition of computer literacy in order to set our goals for computer education. Before embarking on that definition, let us first consider a precise definition of literacy itself. "That person is literate who, in a language he speaks, can read with understanding anything he would have understood if it had been spoken to him; and can write, so that it can be read, anything that he can say" (Gudschinsky 1976, 3). It is not easy to provide a definition of computer literacy with the same precision, nor may it be desirable to do so. Literacy, the ability to read and write, is a rather straightforward skill easily defined. The same is not true of computer literacy because there are so many different aspects of using a computer which require various degrees of skills and aptitudes, from the simple use of the computer to access information to programming a computer to automate a complex information-processing task. Nevertheless, some notion of what we mean by computer literacy is essential and I will therefore attempt to define it in terms of easily describable behavioral outcomes. First, computer literacy is not a discrete state like being alive or pregnant, in which case either you are or you aren't. It is, in fact, possible to be a "little" computer-literate. Computers are ubiquitous and people are more computer-literate than they think. Anyone who has used an automatic elevator has essentially operated a user-friendly menu-driven computer with a very simple keyboard.

Computer literacy is best defined in stages so that it is not a question of whether or not one is computer-literate, but to what degree one is. One does not need to know every aspect of computing to be computer-literate, just as familiarity with all aspects of literature is not required to make one literate or total knowledge of mathematics to make one numerate. So, it is possible to be computer-literate without being able to program or repair a computer, just as one can be a driver without being able to design or repair a car. The stages of computer literacy can be ranked along a continuum. They include the ability to:

1. turn on the computer,
2. access a program on the hard drive,
3. load an application on the hard drive from disks,
4. respond to a menu-driven software package,
5. access information from the computer,
6. input information into the computer,
7. process information using the computer,
8. word process with a computer,
9. do spreadsheet analyses using the computer,
10. operate a database using the computer,
11. create a software routine using an authoring language,
12. program a computer to input, access, and process information,
13. develop a programming language for the computer,
14. develop an operating system for the computer,

15. design a computer,
16. judge when the use of a computer is appropriate, and
17. understand the issues concerning the deployment of computers in different environments in society.

The skills constituting this definition of computer literacy have been ordered more or less with increasing complexity except for the last two points. My definition of literacy does not differ greatly from that of Joel Schwartz:

> Computer literacy takes shape as a group of behavioral skills which permits individuals to utilize computers within the parameters of societal expectations. I suggest that any computer literacy program should teach students the functions, applications, and implications of computer usage. Computer literacy training should, at minimum, promote the following outcomes: 1. promote a basic understanding of how a computer works; 2. provide basic skills for interacting with a computer to access stored information; 3. familiarize students with various applications of available software programs; 4. provide basic skills for using computers to run available software; and 5. develop awareness of the computer's impact on society. (Schwartz 1989, 161)

Computer literacy as I have defined it does not necessarily entail the ability to program or design a computer, just as literacy does not necessarily entail the ability to invent an alphabet or some other new writing system. Schwartz provides two reasons that programming need not be included in a basic computer literacy program. "First, specific programming skills can be viewed as vocational training as opposed to literacy training. Secondly, research findings on the effects of computer programming indicate very little lateral transfer to a student's analytic or problem solving skills" (Schwartz 1989, 161).

A thorough definition of computer literacy, on the other hand, is not limited to the skills of just operating or programming a computer; it includes an understanding of the social implications of the new technology. Information technology is becoming more and more pervasive in modern society. It is "unlike any previous tool. The material it shapes is the most fundamental to human action: information . . . Its impact cannot be studied in isolation from the social context . . . Education must therefore look beyond the walls of its schools and examine the social context itself and see how and in whose interests information technology is already being used, and promoted" (Schostak 1988c, 2). In order to study the social context of the use of computers, students will also have to examine the nature of computers and develop a rudimentary understanding of what they can and cannot do with the technology. Some of the issues surrounding artificial intelligence must be addressed, as well.

Understanding the social context of information technology is not limited to computer literacy; it also includes understanding the social context of the other modes of language. Literacy entails understanding how and when writing is used in society and what its implications are. Innis and McLuhan were pioneers in examining the myriad social, political, cultural, and economic effects of writing and other forms of communication. True literacy involves an understanding of, or at least a sensitivity to, these effects. Similarly, numeracy or science literacy entails an understanding of the social, political, cultural, and economic effects of mathematics and science.

The Limits of Artificial Intelligence

My approach to curriculum development or reform would be termed by some a form of computer enthusiasm. I am happy to accept this characterization of my work, but wish to distinguish my form of computer enthusiasm from that of the proponents of artificial intelligence and the LOGO microworld approach.

"Researchers on artificial intelligence, like Papert, see great potential for increased human intelligence through interactions with computers. Computers act as environments for exploring ideas and stimulating higher order thinking [microworlds] . . . Enthusiasm for computers is driven by research on microworlds based on theories of artificial intelligence (AI) . . . It is important to see that the roots of computer enthusiasm lie in the promise of AI" (Olson 1988, 6–7).

While it is true that a school of computer enthusiasm derives from the AI movement, my computer enthusiasm does not. In fact, I reject the main tenets of the strong AI program. I believe the notion that human intelligence can be programmed into a computer is horribly naive and, as a consequence, dangerous. My enthusiasm for computers stems from the same source as my enthusiasm for basic literacy and numeracy. Just as writing and mathematics greatly expanded the world of learning and became an integral part of both general education and vocational training, so too computing will play a similar role in the information age. We should not forget that writing and mathematics, like computers, created their own microworlds, that is, "environments for exploring ideas and stimulating higher order thinking" (Kurland and Kurland, 319–20). Writing and mathematics, however, did not create artificial intelligence; they enhanced human intelligence and allowed it to flower. The same is true of computing. It is not nor will it be a source of artificial intelligence. Rather, it is a powerful tool which will extend and enhance human intelligence.

The danger of the AI movement is that the limited kind of "artificial intelligence" that a computer generates will set the standards for what is an acceptable level of human intelligence, just as Muzak is regarded by some as music. Computer-generated "intelligence" is not human intelligence at all but a simulation of human intelligence, just as a book is able to simulate a person telling us a story or sharing knowledge but cannot answer our questions or interact with us like the author. One of the sources of the notion that human intelligence can be programmed into a computer is our understanding of the behavior of neurons, the individual nerve cells of the brain and the nervous system. Neurons exhibit a simple binary on–off digital logic in that they only fire an electric impulse when a certain threshold of inputs into them is exceeded. The neurons are networked in much the way as the components of the integrated circuits of a computer. Hence the temptation to draw an analogy between human thought and the logical operations of a computer. Where the AI theory breaks down is that not all aspects of the human mind are controlled by the logical operations of the brain's neural nets. The human mind is shaped as much by the endocrine system and endorphins as by neurons and the flow of electrons. Human thought is as much controlled by pleasure, passion, love, morality, beauty, mysticism, and curiosity as by reason and logic. It is difficult, if not impossible, to explain in terms of logic the music of Mozart, the sculpture of Michelangelo, the paintings of Rembrandt, the poetry of Shakespeare, the philosophy of Spinoza, or the moral ideas of the great religions and their prophets.

The notion of artificial intelligence is a product of human hubris and represents an overextension of science. Each time humankind makes a little progress in understanding nature, there are those who jump to the conclusion that they understand the whole picture. The ancient Greek philosophers who invented deductive logic thought they could explain all of nature with their theoretical, rational, reductionist philosophy; they failed to understand the importance of empirical observation and experimentation. The success of Newton in describing in mathematical equations the motion of mechanical systems led to the formulation of a naive mechanistic view of the universe. The enthusiasm of Alexander Pope as expressed in his lines "Nature and Nature's laws lay hid in night; God said, 'Let Newton be and all was light,'" was typical of those who thought that Newton's mechanics was the final word. In fact, Newton held sway for only 200 years until Einstein formulated his theory of relativity in 1905. The supporters of strong AI, high on the success of modern computing, are making the same kind of mistake in oversimplifying a very complex phenomenon.

Modern science has taught us that there exist natural barriers to our understanding of nature. Goedel's theorem states that a mathematical system cannot be both complete and logically consistent. Quantum physics through the Heisenberg uncertainty principle limits the accuracy of measuring both the momentum and position of a particle. The universe changes each time it is observed or measured. Surely if the precise knowledge of the position of an electron interferes with the knowledge of its momentum, there must be some limits on human intelligence understanding itself. Unfortunately, humility has not always been the strongest characteristic of modern scientists.

Artificial leather is not leather but a material that simulates some of the properties of leather. Artificial intelligence is not human intelligence but a form of computer-generated communication which simulates some aspects of human intelligence. A book that is well written is able to simulate human discourse and provide its reader with a form of companionship. Because the book is a medium whose interactivity with the reader is somewhat limited, it is not regarded as a form of artificial intelligence or even as an artificial storyteller. The computer is a much more sophisticated communications technology which, when properly programmed to anticipate the different possible inputs of its users, can simulate some forms of human intelligence. If the author of the software routine is particularly clever in anticipating all possible inputs and is able to script a set of appropriate responses, it is possible that the computer will be able to fool users into believing they are dealing with a disembodied intelligence. It does not take a great deal of skill, for example, to convince many people that their past and future can be read by a psychic. The psychic uses a form of noncomputer-based artificial intelligence to decipher the responses of their subjects to a prescripted set of questions in order to determine facts about their past life and current desires to simulate a reading of their past and future.

Mathematician Alan Turing's definition of AI is that if you cannot distinguish the communication of the computer from a human communication by telex, then the computer response is a form of artificial intelligence. But a human on a telex line is limited in communicating his or her full range of intelligence because of the inability to communicate information that can be conveyed face to face through body language, tone of voice, and facial gestures. If all our interactions were mediated by telex communications, our culture would not be as rich as it is. If our culture were mediated by AI, the same impoverishment of human interactions and spirit would occur.

I believe that the formulation of strong AI or the notion that the computer can duplicate human intelligence stems from the fact that its proponents regard the computer strictly as an information-processing device and ignore that it is also a medium of communication. A book is not a form of human intelligence, but the communication of human intelligence through the artifact of the printed page. Artificial intelligence is not an independent form of intelligence, but the communication of human intelligence through the artifact of a computer, which is a medium for the storage and processing of information. AI dehumanizes human intelligence by confusing the process and the artifact. An AI system is an artifact. An artifact does not contain intelligence, nor is it intelligent. It merely serves as a medium for human intelligence. Reading a book is not the same as meeting the author and discussing the contents of the book. A telex conversation simulates face-to-face discourse, but it does not replace face-to-face meetings. Negotiations conducted strictly by telex would prove extremely difficult if a delicate matter was under consideration. It is difficult to imagine how computer-based AI could evoke those aspects of human intelligence that encompass care, concern, compassion, love, curiosity, or any of the rich panoply of human emotions which make human companionship so valuable.

Computers have been programmed with all the rules of Mozart's music, but they have never been able to compose a piece of music that comes close to evoking the feelings that his music generates. The genius of Mozart was not his use of rules of composition that can be programmed on a computer, but his talent for knowing when to break the rules and how. This intuition came out of his life experiences, his ecstasies, and his sufferings. How do we make a computer experience ecstasy and suffering or teach it how to weep or laugh? Some human actions or reactions can be quantified and replicated with mathematical algorithms, but some are beyond analytic prescription, like those of joy, remorse, tragedy, love, and spirituality. Perhaps the intelligence of a computer hacker or computer nerd can be replicated through AI, but not that of an emotionally active person or an artist. The danger of the proponents of strong AI is that their impoverished view of the human spirit will discourage others from fully realizing their human potential.

8

THE INTERNET IS THE MEDIUM AND THE REINTEGRATION OF WORK AND LEARNING IS THE MESSAGE

The Total Decentralization of Information Flow

Marshall McLuhan long ago predicted the complete decentralization of the flow of information. The Internet is an embodiment of that idea, a worldwide web of computer users and computer networks linked by telephone lines and fiber optic cables. It is a conduit for E-mail. It is an information highway — a network of networks. The Net allows any two participants to communicate with each other via their modems or routers so that they can exchange any information that can be stored on their respective computer systems. Users can surf the Net to gain access to information files which they can download on their computers, including everything from documents and software to graphic images, photographs, music, multimedia presentations, movies, and videos. It is a global client/server system with multiple numbers of servers and millions of clients and is therefore the client/server of client/servers. It is being used to buy and sell goods, hold forums or discussion groups on a wide variety of topics, provide customer support, promote everything from political causes and commercial products to social activities and art forms. Anyone who wants to start a new discussion group or promote something new may do so. The Internet is a basic information utility whose importance is growing daily. It has already achieved a status equal to that of the postal system, the telephone network, or the broadcast system, and even threatens to surpass these venerable and established institutions. Paul Bonnington, the publisher of *Internet World*, asks in the April 1995 issue, "Is the world wide web the fourth media, a technology positioned to take its place with the big three — print, radio, and television — as a mass market means of communication? It's hard to create an argument against it. The web has all of the social, technical, and economic fundamentals which could help it achieve this prominence" (Bonnington 1995, 3).

Originally, the Internet was organized by the American military and designed to provide a backup communications system if the United States lost large blocks of its normal communications system due to nuclear attack by the Russians. As a result, the Internet was designed without a central switching system or headquarters. It is a distributed network where each of the computers on the Internet contributes to its communication and distribution capability. The Internet makes use of the infrastructure of the telephone companies but it is not operated by any telephone company or other form of central administration. It is the world's largest self-policing community of individuals and organizations operating without any centralized regulating authority.

One of the first nonmilitary uses of the Internet was made by academics, primarily scientists, who used it to facilitate their research activities, mainly through the use of e-mail and the transfer of data and software files from one computer to another. Scientific research has always been a cooperative activity in which scientists have shared their data and their results freely. The Internet is therefore an ideal medium for conducting scientific and other scholarly research activities. In recent years, with the popularization of microcomputing, the Net has also been discovered by the business world as well as by home users and hobbyists. The scientists' ethos of sharing ideas, information, and data, however, still dominates the use of the Internet by other communities and gives it its unique character as an open medium of communication.

The Net boasts approximately twenty million users connected by over 48,000 different networks (*Business Week*, Nov. 14, 1994) and is growing at the rate of 8 percent per month (*Information Week*, Jan. 9, 1995); other estimates of the number of users are as high as forty million (*Toronto Star*, Jan. 22, 1995, E2). The Net straddles international borders and links continents with users in every country and every major city in the world.

It is used by children and adults from all walks of life — students, teachers, workers, business people. It is used everywhere — in homes, schools, offices, businesses, government, research organizations, libraries, and community centers. It links workers to their colleagues, clients, and suppliers. It connects friends and complete strangers. It provides low-cost and efficient communications and provides a medium for information gathering, information dissemination, research, advertising, cataloging, buying, selling, instructing, teaching, learning, congregating, community development, planning, committee work, socializing, entertaining, amusement, killing time, and saving time. It is a place where people meet to play and work.

One of the themes of this book has been that computers contribute to the reintegration of work and learning. The Internet may well be the ideal medium for the reconciliation of these two domains that grew apart with the advent of literacy and the demise of tribal society. McLuhan predicted that electric technologies would retrieve the patterns of oral societies including the reintegration of work and learning. "Paid learning is already becoming both the dominant employment and the source of new wealth in our society" (McLuhan 1964, 351).

That the fragmentation of work and learning is coming to an end is no longer in doubt. As we have seen, it is an idea embraced by Toffler, Senge, Tapscott and Caston, Hammer and Champy, and Drucker. It is a matter of "electronic manifest destiny." Whether or not the Internet will be the actual medium where this reintegration takes place is the focus of our study in this concluding chapter. The merger of work and learning in the computer age represents a reintegration, because in oral societies there was no division between work and learning. There were no schools, no formal education, only apprenticeship or on-the-job training as was described earlier.

The reason that the Internet will play a role in the reintegration of work and learning is that it replicates the patterns of oral society. Many authors evoke McLuhan's notion of the global village in describing the Internet user community, but their analysis usually goes no further than this observation by Internet expert Mary Cronin:

> "Global village" has been applied to many different situations over the past few decades, some far afield from Marshall McLuhan's original concept of how computerized communication changes society. But it seems entirely appropriate that this phrase should become the description of choice for the Internet. The millions of people linked to the network not only have the common experience of using the same electronic tools and information resources regardless of location, they are also connected to each other in a new way, free to exchange ideas, solicit advice, consult, argue, without regard to national boundaries. (Cronin 1994, 64)

THE INTERNET: THE SILENT ORAL TRADITION

The above excerpt from Cronin's *Doing Business on the Internet* is the only mention she makes of McLuhan in her book. It does not do justice to the parallels that can be drawn between Internet usage and the patterns of communication in an oral society. The patterns of oral culture are played out every day on the Internet. McLuhan observed the reversal of mechanical forms into the pattern of oral culture with the introduction of electricity. The microcomputer has reinforced and strengthened the patterns of oral society and the Internet is poised to extend that reversal even further. In fact, one may claim, as many have, that the Internet is a global village.

There are many parallels between the Internet and oral culture. Oral culture was an extension of the psyche of tribal humankind within its community just as the Internet is an extension of its users — an extension, however, with a global reach. The protocols and rituals associated with the use of the Internet are not the formal patterns characteristic of literate communications but are more like those of an oral society, despite the underlying literate substrata that infuses the use of computers. The primary mode of communication, e-mail, is written, but, in contrast to traditional literacy, the form of writing is not formal. Grammatical structures are frequently relaxed and shorthands and jargons are liberally used. The writing is frequently infused with hieroglyphic signs used to connote feelings and tone and are meant to replicate the kind of information that facial gesture and vocal tone convey during face-to-face conversation. Because "wry wit can frequently come across as obnoxious and offensive, humor or irony can be flagged with a 'smiley,' a sideways smile that looks like :-) or :) or a 'grin' that looks like <g>" (Reznick and Taylor 1994, 154).

In addition to communicating text files to one another, Internet users can send graphics, video, music, or any other form of information that they can store on their computers. The medium of information transfer is therefore not strictly textual. But even given that the majority of transmissions are textual, the medium is still an oral one because the pattern in which the information is transmitted encourages dialogue and the other features of oral tradition. The medium is the message and the medium of text-based messages transmitted on the Net is an oral medium even though the content of the messages is not being voiced but is transmitted electronically as text.

DISCUSSION GROUPS ON THE INTERNET

To illustrate the point that the Net is an oral medium, let us compare three types of discussion groups, those of an oral society, those of a literate society, and those held electronically on the Internet. In a literate society, except for informal discussions among friends, there is always a moderator or chairperson who leads the discussion and polices the interaction of the participants, whether the discussion is a political meeting or an academic seminar. In both oral society and Internet society, no such structures exist. There is no hierarchy; each participant has an equal voice and is heard. In oral society, discussions do not have the same kind of precise beginning, middle, and end structure that discussions in literate society do, where there is almost always a precise time frame set for a discussion. In oral society, the discussion goes on until the issue being discussed has been dealt with to everyone's satisfaction. There is no such thing as a member of the discussion group calling for closure and asking for a vote to determine if the meeting can be terminated. I have vivid recollections of intense battles in academic forums in which the topic of debate became whether or not to adjourn the meeting. On the Net, there is no end to a discussion; it goes on as long as any one of the participants is willing to participate. In fact, discussions on the Net are not even limited by the biological necessities that limited discussions within the oral tradition when the participants became exhausted and needed to sleep or eat. Councils were often extended in oral culture by the serving of food during the discourse (the origin of the

Symposium, the Passover seder meal, and the banquet) or by having the participants camp together so that they could continue their deliberations after refreshing themselves with a night's sleep. A powwow or council could last several days, but eventually the participants would have to return to their hunting and gathering activities to find the food necessary for their daily survival. On the Net, discussions are frozen on the hard drives of the participants and the dialogue entered into at convenient intervals. There is never a biological reason to terminate the discussion group. The only factor that can end a session is lack of interest.

NETOCRACY: ULTIMATE PARTICIPATORY DEMOCRACY

In a council in the oral tradition, once a speaker has begun an intervention, he or she is allowed to continue uninterrupted. After the speech, there is a period of silence so that the audience can think about what was said. After a short silence, which can seem like an eternity to a literate individual, someone else will begin to speak. In literate society, on the other hand, people are constantly interrupting one another, vying for air time. This kind of behavior is considered boorish by native peoples, who have retained the patterns and rhythms of their oral cultures even after becoming fully literate. This is not to say there is no dispute within oral cultures; indeed, there is no shortage of that. The pattern of dialogue is completely different from that of literate culture, however. And the dialogue of "the Net people" is not unlike that of tribal people. Net people have their disputes. They may flame each other — that is, send a violently argumentative reply — but they are forced by the technology not to interrupt one another's interventions, so that the voice of each participant is heard. Both oral and Net culture practice a form of total direct participatory democracy as opposed to the representative form of democracy practiced in literate societies, or its complete absence, as is the case in totalitarian societies. It is interesting to note that absolute monarchies, tyrannies, and totalitarian regimes are a product solely of literate cultures. The triumph of democracies that modern literate societies celebrate is nothing more than a celebration of the restoration of the democratic practices of preliterate societies, although literate cultures labor under the myth that they invented democracy.

Totalitarian regimes cannot exist on the Net. The Internet, in fact, is a safeguard of democracy because it cannot be controlled by the state or the wealthy, in contrast to the press or the broadcast media. You cannot buy the Internet or censor it as you can a newspaper. If "freedom of the press belongs to those that own one," freedom of the Internet belongs to those who have access to a computer and can afford an account with an access provider. The appearance of the Freenet in many cities across North America is opening the doors to a wider and more democratic participation in Netocracy. As the cost of computing continues to spiral downward, the economic barriers against participation will dwindle. But cost is probably less of a barrier than interest. A color television represents a sizable investment, yet its use is universal in both the affluent and the developing world. Only *extreme* poverty has presented a barrier to television usage, and the same will eventually be true of the microcomputer and the Internet. The percentage of use of interactive information technology, however, will always be less than that of television and will probably constitute the fault line between the working class and the computerate class (the electronic middle class).

ELECTRONIC CRIME AND PUNISHMENT

In the oral tradition, when a member of society seriously offends the social norms, they are not imprisoned or punished, as is the case in literate society; they are banished: shut out rather than shut in. The same is the case in Networld. If a user violates the rules and protocols of Internet usage, they are first admonished and warned, usually in the form of a flaming, but if their offense continues, they are banned from participation and removed from the discussion group and their e-mail is ignored.

TRIBAL SOCIETY AND RESOURCE SHARING ON THE NET

There is a much greater propensity in tribal society as compared with literate society to share resources than to accumulate wealth. I have attributed this to the fact that wealth could not be stored in a tribal hunting and gathering society as it can in our culture, so that the strategy of sharing maximizes an individual's or a family's chances of survival. In the computer age, information is not only the principal commodity of exchange, but its very nature has changed. Information is no longer private in an electronically configured environment of information. The concept of proprietary information, a holdover from the industrial age, is beginning to weaken and is being replaced with the idea of shared information and open systems. The Internet embraces this new notion. The operating system upon which it is based is UNIX, which was originally developed at Bell Laboratories and is freely available to anyone who wishes to use it — in strong contrast to the operating systems developed by the Apple Corporation and Microsoft. The philosophy underlying the founding of the Internet was the idea that information cannot be hoarded or it becomes useless.

This philosophy parallels that of the sharing of food in tribal societies. The hoarding of food was deemed useless because the food would spoil. Furthermore, sharing it with friends and neighbors put one in a position to receive their surplus in their time of plenty. By and large, users of the Internet have adopted a similar credo. They are willing to share information because they realize that they will be greatly enriched by the process. When information is shared among a group of individuals, their efforts are multiplied. Ten people working together and sharing their information will generate at least ten times as much useful information as an individual working alone. And given that the Internet connects an individual with all its users, the circle of ten can expand potentially to twenty million users. The power of the Internet is not that it puts one in touch with twenty million other users — the network of telephone users is much greater, for example. The power of the Internet is that built into it are tools allowing individuals with similar interests to make contact with one another and share information. Alternatively, once individuals with a common interest make contact through some other medium, they then have a medium through which they can collaborate and cultivate a network of others with a similar interest.

In the course of writing this chapter, I received a telephone call from Jeff Long, director of the Notational Engineering Lab at George Washington University. He had read my previous book, *The Alphabet Effect*, and wanted to invite me to his lab to give a talk and begin a research collaboration. Because we are both Internet users, we did not have to wait the three months until I came to visit him to begin our work together. We began our collaboration immediately by exchanging papers and commenting on each other's work. The distance between Toronto and Washington, D.C., magically melted away. I invite my readers who wish to discuss the ideas in this book or comment on them to do something similar and join the Fifth Language discussion group which I have organized. For those interested, please contact me at Logan@physics.utoronto.ca. I will create a distribution list and forward the comments of others to you.

The open sharing of information has been the strategy behind scientific research from its inception and is what distinguished it from engineering, where information was often kept secret so as to realize a commercial advantage over potential competitors. Science as a language is built on the foundations of literacy and numeracy, yet it has retained many elements of the oral tradition. The first users of the Internet were scientists who infused the current-day Net with its cultural traditions of open discussions and the full disclosure of information. With the Internet, the circle of research collaborators becomes global. Science has always been a global activity, but with the Internet and other electronic links, collaborations can be carried out in real time because of the instantaneous nature of communications. With the institution of open discussion groups on the Net, scientific activities have become even more democratic and inclusive. Lack of academic credentials have sometimes stood in the way of scholars who had significant contributions to make. This is no longer the case with electronic discussion groups where ideas are often examined on their own merit, stripped away from the status or expertise of their author.

One reason that information is easy to share on the Internet is that it is so easily multiplied and communicated to others. There is no cost involved in its distribution on the Internet. Users are, therefore, eager to share and participate in a commerce that enriches all who participate. There is really no such thing as a bad deal when information is being shared. No one ever feels cheated, as is sometimes the case when material goods are exchanged.

THE INTERNET: BARD AND STORYTELLER

In tribal society, the users of information are also its creators. In the tribal council, every voice is heard and every pair of ears is part of the system. Even when the storyteller told his or her tale, it would change depending on the mood and reception of the audience, as was pointed out by Eric Havelock (Havelock 1963). Internet participants can do more than just shape the story by indicating their mood; they can actually change it by adding their comments. The story being told over time becomes their story, a never-ending story. They feel a sense of ownership. Internet participants are expected to participate in the flow of information once they have had time to get the drift of the discussion. "Newbies," new participants, can follow the discussion for a while without contributing, but if they continue to be passive participants, they are labeled "lurkers," which is a pejorative for an electronic Peeping Tom or voyeur.

One of the most interesting developments on the Internet has been the emergence of web pages through graphical user interfaces such as Mosaic or Netscape. Web sites are the corporate bards of the Internet. A web site is a location on the World Wide Web where Internet users can access information sitting on a host computer. The web site is accessed through a home page to which the other web pages may be reached through hypertext links. This structure allows users to access the information that best matches their interests and information needs, and to linger over material of particular interest or skip over irrelevant material. Linked pages using hypertext retrieves the patterns of the oral storytellers. Hypertext leads from one story to another, depending on the interest of the user. In the oral tradition, the storyteller told one story after another, depending on the interest of the audience. There was no fixed order to how the episodes were related to the audience. Each performance was different and the episodes were told in a different order each time (Havelock 1963). Hypertext is like jazz. Each time the story is told, it is told as an improvisation, one dictated by the users and their needs and not by the producers or source of the information.

The fragmentation inherent in many of the communications and information-processing patterns of today's institutions are holdovers from the industrial age. These forms of fragmentation will begin to disappear as the Net reintroduces the integrative patterns of oral society. Generalists are able to participate in an area of interest to them on the Net without the prerequisite of the specialist knowledge that is required for entry into academic discussions. Patterns of usage on the Internet are:

1. oral and tactile rather than visual and literate,
2. right-hemisphere rather than left-hemisphere,
3. synthetic rather than analytic,
4. nonlinear rather than linear,
5. simultaneous or synchronous rather than sequential, and
6. analogical rather than deductive.

THE HISTORY OF PREELECTRONIC PROTO-INTERNETS

The discussion group on the Internet can be likened to an electronic dialogue or Symposium similar to the one described by Plato in his Dialogues, but one in which the participants share a common corner of cyberspace rather than sit at a banquet table. The Internet is obviously not the first forum for the sharing of information. The roots of the Internet or information-sharing networks can be traced to the preliterate storytellers or bards who traveled from settlement to settlement sharing information, carrying news and relating stories from the past which formed the foundations of their culture. Next came the forums of the city-states like those of Babylon which acted as nodes for receiving and sending information. The subsequent manifestation was the imperial communications systems of Mesopotamia, Egypt, Hellenistic Greece, the Roman Empire, and the Roman Catholic Church. In each of these proto-Internets, communications were controlled by and emanated from a central authority (all roads lead to Rome). There was heavy policing of the information transmitted through these systems so that only the Imperial political line or the Church dogma received air time. While these closed systems were opposite in spirit from the openness of today's Internet, they did provide global coverage (at least empirewide coverage) and, relatively speaking, simultaneity. The ancient Roman system of roads and the use of paper documents with a response time of only a month or two to the farthest outposts of the empire provided the fastest distribution of information up to the time of the introduction of the railroad and the telegraph.

The proto-Internets of the Renaissance and early industrial age were the trade routes along which information was carried by traders and merchants along with their goods. These routes became truly global with the arrival of ocean-traveling ships which permitted European powers like England, France, Spain, Holland, and Portugal to set up global trading networks that gathered goods and information from the four corners of the world. The newspapers that emerged in these countries provided readers with news of events from faroff places. With the invention of the telegraph, newspapers were transformed into a global network of correspondents, with the metropolitan newspapers like the *London Times*, the *New York Times*, and *Le Monde* serving as nodes. The link between these nodes where information was collected and disseminated was the international telegraph system, whereas the link to the end users were the daily newspapers, which provided information that was no more than twenty-four hours old. The next proto-Internets to arise were the broadcasting networks that were spawned first by radio and then by television. The feature that all of these proto-Internets lacked was the interactivity of end users, who functioned merely as passive consumers of information.

The Electronification of Information

As McLuhan foresaw, the electrification and then the electronification of information resulted in new patterns of information usage that resemble the oral tradition in many ways. This trend, which began with the introduction of the telegraph and continued with the use of the telephone, radio, and television, is finding its most complete actualization with microcomputers connected through the Internet. Paradoxically, the bulk of the information being transmitted on the Net is still text. Print as a medium is still holding its own. The sale of books and other forms of reading matter continues to increase. The information which is now collected in books has been gathered, written, edited, and typeset using electronic media, but the format of print on paper still seems to be the most sensible format for reading text. Only the briefest messages, such as e-mail, are read exclusively on a video screen. When lengthy files are transmitted from one computer to another, in most cases readers will make a hard copy of the document. The reason is simple; reading from a CRT (or video screen) is an unnatural activity because of the way the brain processes video information. Reading is a left-brain activity, whereas viewing video is a right-brain one. The mosaic pattern of light pulses must be reassembled by the right brain to create an image. There is an inherent conflict in reading directly from a CRT. This is why the book is still a popular medium and will continue to be so. The Internet will not replace the book when it comes to the distribution of lengthy literary material, but it will remain the medium of choice for short text messages and multimedia formats.

Paper-based journals that publish specialized information could easily disappear, however. As archives for information and knowledge, they have long been surpassed by electronic media, and with the Internet they have also been surpassed as distribution vehicles. Rather than printing all its articles on paper and mailing them to their subscribers, journals could distribute their material electronically, leaving it to readers to print out those articles they wish to read closely in hard copy. But what purpose would electronic journals serve if this were done? Why not just have authors distribute their articles individually over the Internet? While this does happen, there is still a reason to have a journal, even if it is in an electronic form. First, the journal serves as a means of creating a community consisting of its readers and its contributors. Second, the function of editing and vetting papers by peers can still be maintained. Disputes between editors and authors can be easily resolved by distributing both authorized and nonauthorized articles. Readers would enjoy the service of having articles submitted for publication vetted by an editorial board and yet have the option of reading or ignoring those articles which were not officially approved by the board for formal publication.

The ease with which text can be read is critical to the success of a medium. This principle is illustrated by the different fates of two media which were introduced at approximately the same time, namely, videotext and desktop publishing. Videotext involved the broadcast of information into the homes of its users, who were able to select what they would view, paying a small fee for each item selected. The service never got off the ground in the U.K. and North America, partly because there was little context to the information provided and also because the video medium was not conducive to the easy reading of large amounts of data. The fact that information could be kept up to date electronically and was therefore always more current than printed information could not overcome the medium's inherent disadvantages. Another drawback of videotext services compared with the Internet was that the users of videotext couldn't make a contribution to the information hoard being compiled and so felt no sense of ownership of the information. The Internet user, on the other hand, can interact with the on-line information by adding to it or commenting on it for other users to read. Thus, users are able to shape the overall direction of a newsgroup in which they are participating and hence develop a sense of ownership of the information, even if it is shared with other users.

The desktop-publishing medium based on the rapid and relatively inexpensive production of printed material using the laser printer and software packages like PageMaker or Quark has caused a revolution in the print business. Traditional typesetters who could not adjust to the new medium and learn microcomputer-based typesetting were thrown out of work, and printing houses that did not adopt this new technology soon found themselves out of business. Desktop publishing takes advantage of the medium of the laser printer, the electronic image setter, and the microcomputer to reduce the cost of production, make short print runs economical, and increase the frequency with which catalogs and newsletters can be updated. At the same time, it retains the advantages of the crispness of the printed page. Material can be copied directly from laser-printer output or sent to an image setter to produce film for phototypesetting. It is even possible now, using computer-to-plate technology, to bypass the film stage and have the computer directly etch the plate.

THE INTERNET, CD-ROM TECHNOLOGY, AND MULTIMEDIA

The Internet is no longer restricted to text; it is rapidly becoming an important vehicle for the transmission of interactive multimedia data. Multimedia, which combines text, visuals, and audio, owes its current success to CD-ROM, which has served as both a storage medium and a delivery device for this interactive medium. At one time, the CD-ROM medium totally monopolized the distribution of multimedia products. With the Internet and powerful servers, this is no longer the case. As the bandwidth of modems increase, allowing more data to pass from one user of the Internet to another, multimedia presentations will become more practical. The Internet will have a major advantage over CD-ROM. The information on a CD-ROM is frozen and cannot be updated, whereas Internet-delivered multimedia information can be updated dynamically. By coupling a web page server with a database server, web pages can be updated dynamically so that the information being delivered is never out of date. Multimedia presentations on the Internet are, as of the time of this writing, primitive, but the capability for more sophisticated material is there; it is only a question of time before the Internet fully supports multimedia.

From the advent of writing on clay tablets to today's Internet, there has been a continuous speedup of the transmission of information. Harold Innis (Innis 1951) has documented how the transition from clay to papyrus and then paper speeded up the transmission of information, as did the introduction of the printing press. Word processing on computers, followed by desktop publishing, increased the speed with which information could be transmitted and the ease with which it could be edited, changed, and updated. CD-ROM promised to extend this speedy and flexible capability for both text and nontext information such as graphics, sound, and music. CD-ROM has become standard equipment on most microcomputers manufactured today. Despite this, CD-ROM could be squeezed out as the medium for the active transmission of multimedia by the Internet and survive only as an archiving device. My guess is that canned packages like entertainment and educational titles will find a natural home in the CD-ROM environment and will be produced and distributed much like books, audio recordings, and videotapes of movies. Information that must be kept current, like commercial information and catalogs, corporate annual reports, advertising in the form of infomercials, will likely gravitate to the Internet or other networks like Compuserve, Prodigy, and America On Line. With the recent appearance of recordable CDs, the dynamics could change once again, with recordable CDs being distributed with the most current information, which could be regularly updated through links on the Internet. As in the past, the most likely scenario is some hybrid system making use of the best of Internet and CD technologies.

The Information Superhighway

As computers become more versatile at handling different information formats such as video, graphics, and audio, and as modems or direct digital lines like ISDN become capable of handling greater amounts of bandwidth or data, the Internet will become more and more multimedia and will eventually evolve into the information superhighway that has been the focus of so much recent speculation. The term "information superhighway" is a useful one because it conjures up the image of information traveling quickly from one point to another. The metaphor breaks down, however: a superhighway is a linear system with controlled access, whereas the Internet is a web in which access is not controlled. There has been so much hype surrounding the term that we are stuck with its use whether it is accurate or not.

Terminology aside, the reason I believe that the Internet will become the information superhighway is that with at least twenty million users and the capability to integrate any other information service into its web, it has too much of a head start for any other initiative to take over as the premier conduit for global information flow. Its ethos of free exchange of information without the interference of government and the domination of business is too powerful to permit a serious rival to gain control of the uncontrollable, the free flow of information. The Internet encompasses the dream of the earlier developers of microcomputers, many of them hackers, who envisioned the unfettered flow of information. Ironically, some of these early dreamers went on to form some of the largest and perhaps most proprietary and predatory corporations in the microcomputer industry.

The Internet and Commerce

The Internet is rapidly becoming an important business tool. Commerce on the Internet will retrieve the patterns of the oral society of hunting and gathering and early Neolithic culture. Just as the exchange of ideas among academics and hobbyists using the Net is done on the basis of barter, a similar pattern is emerging among commercial users of the Net. Discussion groups often involve the free exchange of ideas among individuals who might ordinarily have a customer–supplier relation in which a fee is charged for the exchange of information. In fact, the most frequent commercial users of the Net are precisely those individuals from the information and communications sectors who make their living from the exchange of ideas and information. The Internet is even retrieving the patterns of cooperation characteristic of oral culture among traditional competitors and ameliorating the cutthroat competition of traditional business. Discussion groups often involve the exchange of information and ideas from workers in companies that are in direct competition with each other. Individuals who participate in such exchanges are more concerned with solving a mutually shared technical or intellectual problem than in securing some advantage for their firms. Their activities, which no doubt benefit society as a whole, have to be a source of concern for the senior managers of their companies. Management is just beginning to realize that the Internet is a difficult force to control. Like the microcomputer that preceded it, the Internet will require a strategic deployment if it is to contribute to the firm's long-term interests.

The Internet is a double-edged sword. At the same time that it is retrieving some of the cooperative patterns of an earlier era when the pace of commerce was considerably slower, it is also accelerating the pace of business by creating the expectation that information can be made available instantaneously. When the postal system broke down and courier systems came into popular use, there was a general speedup of information flow. A similar speedup occurred with the introduction of the fax machine and then the fax modem card that converted desktop computers into fax transmission centers. With the Internet, the speedup is even greater, with direct desktop-to-desktop communication with no direct cost associated with the transmission of information over long distances. The Internet provides instantaneous communication right across the globe. Space and physical gaps literally disappear on the Internet, as does the time for the transmission of information. The Internet is therefore driving business process reengineering even harder and making the time wasted in unnecessary handoffs totally intolerable. It is also integrating a number of business processes which will require realignment.

The Internet is gaining acceptance in the business world at a phenomenally rapid rate reminiscent of how microcomputers took hold in the early eighties. Like microcomputers, the Internet is being introduced at the grass-roots level in a random manner by individual employees who have discovered that they can enhance their performance at work by using this communication channel. The same chaos that followed the introduction of micros is about to repeat itself with the Internet. Even in those cases where the Internet has been introduced companywide, there are no Internet strategies in place. Companies that adopt the Internet do so largely to realize the cost savings in communications that the use of this technology affords, rather than to realize some overarching organizational objective.

It should also come as no surprise that the technology is being used principally to perform tasks that were carried out previously by computers, telephones, fax machines, courier services, and the post office. The content of any new medium is some previous medium and this is certainly the case for the Net. The traditional tasks being performed on the Net, however, are already being colored by the Net's information-sharing ethos. This has led to much more open discussions and the sharing of experiences than has been experienced with traditional media. One of the most interesting phenomena are electronic discussion groups. These groups focus on a topic which can range from an area of research or professional activity, to a shared interest such as a musical form, a hobby, a period of history, or the use of a product, whether it is an automobile or a software package. Discussion groups perform a function similar to a number of different types of organizations, such as a professional association, seminar group, research team, club organized around a shared interest, and so on. The difference is that the Internet discussion group is in continuous session with no geographical barriers and members can participate at their convenience.

The value of the Internet to business is just beginning to dawn on the community as a whole now that the pioneers have demonstrated its many uses for commerce. Because the orientation of businesses, whether they are members of the information, service, or manufacturing sector, is the provision of a greater level of service, communications is one of the key ingredients for commercial success today. The Internet, which automatically combines communications and computing, is becoming an important tool in contributing to the creative use of communications in business. Some of the uses of the Net include the cost-effective exchange of e-mail and the extensions of wide area networks (WANs) to include isolated co-workers who operate small satellite offices. The Net is used to track standards, conduct technology watches, gather badly needed information, obtain answers to questions, disseminate information, promote products, support customers' use of services and/or products, send and receive all types of documents, and obtain software codes to reduce the cost and time to develop systems software. The Net is particularly useful for conducting customer relations, market research, and product support as well as obtaining market feedback.

Analysts often make the mistake of judging the value of the Internet solely in terms of its contribution to direct sales. The true criterion for judging its usefulness should be strictly in terms of whether or not it helps an organization achieve its goals. The Internet is probably much stronger as a communications, marketing, and customer-support tool than as a sales tool, but by fulfilling these other roles it can increase sales if its use is guided by an Internet strategy which is in synch with the company's overall objectives. With a strategy, other benefits, such as the more efficient operations of the different functions of the organization, will also ensue. For instance, companies spend too much time generating information internally and do not take advantage of existing information that they could gather on the Net. The Net has to be used wisely, however. It is like a dragnet which pulls everything out of the sea, and unless a strategy is in place to filter the data collected, time and effort will be wasted.

The Internet is an excellent marketing research tool. "Well informed employees can spot marketing opportunities, the emergence of new competition, unmet customer needs . . . but only if the company has organized its internal information sharing structure to incorporate their insights" (Cronin 1994, 17). With an Internet strategy, firms will be able to take full advantage of the Net. But the strategic use of the Internet will require the cooperation and coordination of the whole organization and hence will require the reengineering of key business processes, especially those concerned with marketing and customer support, as well as a modest amount of personnel reorganization and training. All levels and aspects of a company's organization can benefit from the use of the Internet.

One cannot expect individual members of a firm, however, to spontaneously optimize the organization's use of the Internet by pursuing their own particular interests willy-nilly. Leadership must come from a broadly based group of managers with experience in both marketing and information technology. Responsibility cannot be relegated to an MIS department which is unable to develop a strategic direction for sales, marketing, and customer support. As Harvard business professor Michael Porter (Porter and Millar 1991) points out, senior managers and owners of a company can seriously fulfill their managerial responsibilities without becoming directly involved in the management of the information technology in their organization. A champion of the technology within an organization who is not in the position of the decision maker is well advised to find a way to get members of senior management to sit down in front of a computer and discover for themselves the power of the Internet. Once they use it and explore its possibilities, senior managers will discover what a powerful tool it is. A modest investment in time and training is required to become familiar with the Net, but it is well worth it, as any regular user will attest.

The Internet is considered a luxury item which provides a modest competitive edge to some companies that have particular information-dissemination needs, such as a software company or a chemical company whose customers require technical information to use the company's products. It will not be very long, however, before its use will be considered as indispensable to doing business as the telephone, the computer, or the fax machine. In fact, there are already firms that will no longer do business with another firm unless they are connected to the Internet. The Internet is particularly valuable for small- to medium-sized organizations that want to support a worldwide customer base.

Marketing is probably the most fertile field for use of the Net. The first instinct is to use the Net as an advertising medium, that is, using the old medium of television as the content for the new medium of the Internet. This approach is doomed to failure because it does not take advantage of the interactivity that the Net offers. Traditional use of media for display ads in print media and spots in broadcast media have lost their effectiveness. After almost fifty years of being bombarded by rapid-fire sound bites and flashy images, the public has become fairly immune to the effects of traditional advertising. To have an impact today, a superficial hit will not work; it is necessary to *educate* one's potential customers about the advantages of the product or service one is offering.

This is why marketing using the Internet and its potential for interactivity can be such a powerful strategy. By providing information with which potential customers can interact, companies can allow individuals to learn about a product or service on their own terms without feeling that they are being manipulated. Customers can and will sell themselves on the advantages of a company's products or services on the basis of this information. By providing such information as product announcements, newsletters, frequently asked questions (FAQs), corporate profiles, pricing, product descriptions, and so on, in an interactive format, one is basically narrowcasting to potential customers and at the same time unobtrusively making a large amount of data available to them, allowing them to customize the information to suit their needs and interests. Because the Internet is already segmented into micromarkets, it is easier to narrowcast one's message. The challenge of using the Internet for marketing is to get people to visit one's web site. One cannot rely on the Internet itself to do this job because the ethos of the Net frowns upon direct advertising and solicitation. Instead, one must deploy other, more traditional information channels to make potential users aware of one's web site. Having a telephone is an indispensable part of running a business, but one does not rely on the telephone to inform potential clients of one's telephone number. The same applies to the Internet. If one creates an interesting site that has information of relevance to some audience, however, word will soon spread and people will flock to the site.

ALIGNMENT: A STRATEGY FOR CONNECTIVITY AND THE USE OF THE INTERNET

The Internet provides an organization with the potential for total connectivity with all of its customers and suppliers and hence permits the integration of a number of its key activities, including marketing, advertising, sales, order fulfillment, distribution, customer support, market research, market feedback, and product design. For these activities to work well together, they must be coordinated and aligned so that they are working together within a strategy to achieve the overall objective of the organization. The alignment team must therefore consist of experts in marketing, communications, and information technology.

Alignment goes beyond business process reengineering (BPR) and takes the process of integration represented by BPR a step further. BPR redesigns business processes so that the efficiencies of using computers can be maximized. Alignment takes those reengineered processes and aligns them, to maximize the flow of information that arises from the connectivity of the Internet, electronic data interchange (EDI), or any other network that connects the firm to its market and suppliers. Alignment is to the Internet and connectivity what business process engineering is to computing.

When the microcomputer was introduced, it allowed working groups within an organization to pursue their own information-processing strategies to fulfill the objectives of their particular division. Work groups often found themselves in conflict with other working groups and with their MIS department, which lost control of the information-processing activity within their organization. With client/server systems and BPR, a measure of order and control has been restored so that, by and large, information technology now serves the overall objectives of organizations.

Because of the volatile way in which the Internet is penetrating the business world, connectivity, like microcomputing when it was introduced, is poised to once again disrupt the information and communication functions of many organizations. This is a pattern that has repeated itself each time a new information technology emerges. A period of chaos ensues in which new applications and methodologies are discovered which are often in conflict with the techniques of the past. The content of the new medium is always some older medium. The first applications of the Internet are going to be made to the traditional ways of doing things until the true leverage of the increased connectivity can be fully digested, understood, and integrated into the traditional objectives of the organizations using the Net. At the next stage, the objectives of the organization will begin to change as new capabilities manifest themselves.

The tools that have been developed to facilitate global communications on the World Wide Web are also ideal for setting up communications links between the computer users within an organization. Organizations have a wide variety of interfaces and protocols to retrieve information. The HTML format and HTTP protocol were designed precisely to allow users of disparate formats and protocols to communicate with each other around the globe. HTML is therefore ideal for use within an organization which is linked by a LAN or WAN or even in those cases when no network exists. HTML is the *lingua franca* of networked computer users. It supplements E-mail communications by providing graphical interfaces and takes advantage of hypertext links. By setting up a Corporate Wide Web and an electronic newsletter, the members of an organization will become better informed of each others' activities. This will contribute to the coherency of the organization's outside communications and the alignment of its operations and activities. Individuals can create their own home pages and those of their working group.

The Electronic Academy

The interactivity of the Internet is a feature that contributes to its success as a business tool and at the same time allows it to function as an environment for continuous learning. The Internet functions as an educational medium just by virtue of the fact that its users are exchanging information with one another and interacting actively with that information by reading it and responding to it. The Net is not a medium like television where the user can be a passive recipient of information, unless the Net is being used simply to download graphics and video. Participants in discussion groups develop new skills. They develop the skills of electronic or Internet rhetoric, which combines the skills of classical oral rhetoric with those of literacy. It is possible that new cognitive structures and skills will develop from using the Internet because the patterns of use and searching for information are so different from anything else.

It is too early to determine how the Internet will be used in schooling and training. The Internet is already operating as a medium for distance education. It is being used to deliver on-line courses in which class participation and student–teacher interactions take place over the Net. The contents of these courses are no different from traditional classroom courses. Once again we find that the content of a new medium is that of an older medium, namely, classroom courses. As the Internet is used more and more for on-line courses, its unique format will begin to color these offerings, and within a short period of time, new courses and new forms of education will emerge. Even in those cases where traditional classroom activities are being pursued on the Net, the new medium allows the participation of those who could not attend standard classroom sessions because of considerations of distance, scheduling, or physical handicap. The Internet is already changing the nature of education by providing distance education and allowing students to gather information and dialogue on a global level.

LIFELONG LEARNING: JOB SECURITY IN THE INFORMATION AGE

Thirty years ago, Marshall McLuhan described the process whereby the electrification of information reversed the fragmented patterns of work and learning characteristic of the industrial age and the age of literacy and retrieved many of the holistic patterns of oral society. The Internet embodies the spirit of the integration which McLuhan identified. It literally integrates all forms of media by providing a global networked digital medium that can accept any form of information that can be digitized, from text and visual images to audio and music. Information flows from one domain to another and by so doing breaks down the barriers between work and learning, entertainment and education, work and leisure, technology and culture, art and science. Computing and the Internet represent a new cultural milieu, a new form of economics, a new way to earn a living by learning a living, and a new way to live.

McLuhan predicted that automation would end "jobs in the world of work" (McLuhan 1964, 346). By the end of jobs, McLuhan did not mean that automation spelled the end of work but, rather, a new way of organizing work. Instead of holding a job, he suggested, an individual would pursue a career and thereby assume a role in society. Consider the position of secretary which has, due to the influence of information technology, evolved into the role of administrative assistant. Secretaries once took dictation, typed documents, and served coffee. They worked from nine to five, at which time they put a cover on their typewriters, went home, and forgot about work. Today's administrative assistants, on the other hand, are junior to middle managers. They make decisions, organize activities, and anticipate problems rather than merely react to them. They stay late and work weekends if required. They are more immersed in their work than the secretary of yesteryear. They are more likely to learn more about their industry, take courses in their spare time to upgrade their skills, and network socially with colleagues and associates in their field.

Like many others in the work force, they are not just earning a living by filling in time, they are pursuing a career and "learning" a living.

During the industrial age, it was common for an individual to work for the same company for an entire lifetime. Today, that is almost unheard of. An individual just entering the work force will probably work for ten or more organizations before retiring, if he or she retires at all. Moreover, the company that individuals work for might go through as many changes as their employees because of mergers, acquisitions, strategic alliances, and restructurings. Between jobs, many of today's workers do not become unemployed; instead, they become a consultant or a casual worker. They move in and out of employment or from one company to another, from one project to another, or from one working group to another. The only thing that is constant is change.

One no longer works at a job; one lives a career. So individuals better like what they do because work is no longer a nine-to-five, Monday-to-Friday activity; it is an all-consuming way of life. Also, the nature of a career has changed. It is no longer a linear progression up a hierarchical ladder of increased authority and responsibility based on seniority. Organizations, particularly those in the private sector, are becoming like the Internet, where no one is in control anymore. Senior managers are increasingly frustrated by the fact that they are not always able to steer their organization in the direction they want. The reason for this is that today's organization, like the tribe or village in oral society, is becoming a self-organizing system. A system of command and control no longer works, but inspiration, motivation, and leadership do. And that leadership and motivation do not necessarily have to come from the top but from anywhere in the organization.

In an environment where change is pervasive — technological, economic, structural, and organizational — there is no such thing as job security; the world is changing too rapidly. The only form of security is to have a set of skills that are up to date and in demand.

This can only be achieved through lifelong learning, in which individuals are constantly honing and broadening their skills through reading, training courses, and intensive study. The Internet is a useful ally for those who want to keep abreast because it facilitates regular monitoring of trends on a global scale and allows easy and instantaneous personal contact with those pursuing a similar career. The only defense against the rapid rate of change that characterizes our times is to try to anticipate the change that is about to take place and then set about to acquire the knowledge and skills that will be needed to cope with the new conditions, not only for today but for tomorrow and the day after, as well.

THE EMERGENCE OF THE COMPUTERATE CLASS

Just as literacy created a new social class and a new form of privilege and economic opportunity, the use of computers may do the same. Those who will be able to exploit the power of computers will have an important advantage over those who cannot, just as those who were literate had an advantage over those who were not. I predict that the middle class will divide into two subsets, one merely literate and the other both literate and computer literate.

The new social class that will emerge, the computerate class, will be able to realize a commercial and cultural advantage. The consequence will be extremely important for educational planners. Just as the emergence of literacy as a skill affected the patterns of education in literate societies, so, too, will the emergence of computers and the need for computer literacy create major changes in and new demands upon our current education system.

Computing also seems to be restructuring the way in which the existing classes relate to one another and changing the role that class structures play in the economics and politics of society as a whole. Computers allow the integration of many different functions and make each individual worker more self-sufficient. There is also a parallel merger of the three classes. Many of the odious and brute-force tasks of the working class are being performed by robots and other computer-controlled machinery. The percentage of jobs that are unskilled or strictly working-class continues to decrease, while the number of middle-class jobs requiring education, literacy, and computer skills increases. Another effect of computing on the breakdown of class structures is that senior and middle managers are less and less dependent upon support staff.

At the other end of the spectrum, it is easier for middle-class people to become masters of their destiny and enterprise. Information-age enterprises are much less capital-intensive than industrial-era ones. Often, all that is required is intellectual capital and a modest amount of cash capital to start a major business operation. There are a number of middle-class computer whizzes who have become multimillionnaires in a relatively short time. Another sector where the middle class is able to become self-employed and often self-realized is the service sector. While work in this sector does not often lead to a glamorous and aristocratic lifestyle, it has for a number of entertainers, fashion designers, and restaurateurs.

The computer age will continue to see a greater merging of the three traditional classes. The control of information, however, will become more and more the key to success and power. It is natural, therefore, that those who are computer literate will begin to realize tremendous advantages. As to whether or not this advantage will lead to the emergence of a distinct class just when class distinctions seem to be on the decrease is impossible to predict. There is no doubt, however, that computers will be where the action is and they will play an ever more important role in education and educational planning.

The Sixth Language?

This study was based on the idea that computing, or the fifth language, represents the current end point in the evolution of language. This notion gives rise naturally to the question of whether language will continue to evolve and if it does what form a sixth language will take. There is no question that language will continue to evolve, but as to the nature of the sixth language, I can only guess. And guessing would be equivalent to attempting to predict the existence and nature of science fifty years after the emergence of writing and mathematics. An impossible task. Nevertheless, let me close with a bit of speculation. If I was forced to guess, my conjecture would be that the jumping-off point for the emergence of a sixth language will entail some combination of general systems theory, business process reengineering, alignment, connectivity (the Internet, EDI, and more), and ubiquitous computing, that is, the increased role of computers in work, education, and everyday life.

APPENDIX
THE COMPUTER AS THE IDEAL CLASSROOM TOOLKIT

This appendix considers the use of specific software applications in the development of school-based curricula for education at the primary- and secondary-school level. As the main objective of the core curriculum being proposed is the development of information-processing skills and computer literacy, I will focus on the basic tool applications of computing. We will examine the application of these tools to the development of communication skills and to science education.

Word Processing and Learning to Write

The most important condition for the successful use of word processing in writing instruction is not the word processor or technology itself but the approach to teaching writing or composition skills. It is essential that the instructor understand that writing is a complex process consisting of many different cognitive activities or elements. Older linear models (Rohman 1965, 106–12) in which writing was regarded as a set of temporally sequenced stages consisting of prewriting, writing, and rewriting are much too simplistic. Studies based on "think-aloud protocols as a writer works" (Pea and Kurland 1986) reveal that writing is an activity in which the writer recursively creates new ideas, plans, drafts text, reviews, revises, and simultaneously monitors all of these activities. In addition to juggling these tasks, authors are composing their text within the context of a variety of constraints, which Pea and Kurland refer to as the rhetorical problem, and are driven by their subject, the genre of the document, their goals in composing the piece, the effects they wish to have on their reader, the tone they wish to develop, and so on. Writing may be regarded as a form of complex problem solving. "To write is to think and reflect; in short, writing is a complex cognitive task, one prone to obscurity because of the many cognitive demands that impinge on the writer at once" (Pea and Kurland 1986).

A successful program of instruction must therefore incorporate the understanding that writing is a complex process as described above, identify the various elements of the process, and create an environment in which students can develop the cognitive skills needed for the process of writing. Within the context of writing-process instruction, word processing can be a very powerful tool for teaching novices how to write and manage the many tasks involved in successful composition.

First, word processing is an excellent medium for carrying out the process of writing because of the way it can integrate the various tasks in the writing process. Secondly, microcomputers, whether they are networked or not, are an excellent environment in which to create a writing community among peers because of the ease with which texts can be shared and because computers seem to encourage multiple social interactions and cooperation within a classroom environment.

The computer helps integrate the flow of information and makes it easily accessible. Once a plan of the structure of a document has been developed, it can be stored as a document separate from the main text and called up at will. The computer facilitates the editing of the text by providing both a hard copy and an electronic version of the text. The hard copy is useful for the purposes of evaluation and revision. Moreover, it is easier to read than the electronic version and it also provides the author with a view of the text as the reader will view it. Comparisons, evaluation, and revisions can be carried out electronically; this has its own distinct advantages because of the ease with which one can move about the document. This process provides a right-brain view of the text, whereas an examination of a hard copy of the text provides a left-brain view.

The computer also facilitates the actual creation of text, that is, the translation of ideas into words. For many users, it turns a tedious task into an activity with some gamelike features, as evidenced by the fact that most students prefer to write with a computer rather than paper and pencil. The computer makes the revision process less tedious because it is so easy to make changes electronically compared with paper and pencil or typescript revisions. Consequently, students revise their texts more often with word processing and write more, as well. Moreover, the computer facilitates the writing process because of the ease with which the author can move from one information activity or domain to another.

STUDENT MOTIVATION AND WORD PROCESSING IN THE CLASSROOM

The history of word processing in the schools is as old as microcomputing in education itself, since word processing is one of the first applications that teachers used in their classrooms when computers were introduced in the late seventies. In the Toronto school board, for example, word processing led to the grass-roots introduction of computers in the schools.

> School F was the first in the Board to begin using computers. This was a result of an initiative taken by one of the school's primary teachers, Mr. Vayda, who was able to persuade the Board to lend him several Pet computers in order to pilot their use. The computers entered the school in the fall of 1978. According to the principal, the computers were used at first for games but it

> was quickly seen that they were capable of much more . . . and the emphasis changed [to] word processing. (Sullivan et al. 1986)

One of the principal reasons for the successful introduction of word processing into the schools, no doubt, has been the enthusiasm with which the innovation has been embraced by both students and teachers and the motivation to write that the medium seems to inspire. "Almost all researchers in word processing and writing note that using a word processor improves writers' attitudes toward writing. Students are so willing to testify to their new enthusiasm for computerized writing that there seems little need to study it systematically" (Barker 1987, 112). What is it about word processing that motivates schoolchildren to write and makes them so enthusiastic about the medium?

Interviews with schoolchildren we observed using computers in the classroom revealed that one of the features that the children liked about computers is that they make written work simpler. In a sample of forty-six pupils who were surveyed or interviewed and who had used computers for word processing, thirty-five indicated that they preferred computers for writing stories, while nine preferred paper and pencil and two had no preference. These results are similar to those of other researchers who have reported that children found writing with computers more fun and easier than using a pencil. Children liked the fact that they did not have to recopy their work each time they made a revision. Consequently, they wrote for longer periods of time and hence produced more text. The behavior of one young boy provides an amusing example of many schoolchildren's enthusiasm for word processing.

> He loves writing stories. He just does it mainly for fun because Tom [the teacher] doesn't let them use the computer for writing their regular writing assignments. He can't save the stories because there's no way to save them. However, as soon as the computer is turned off, his story has disappeared. It's interesting that he finds this satisfying enough to keep right on doing it. He still says that it's one of his favorite activities with the computer,

to write stories. (Sullivan et al. 1986)

One of the more frequent responses of children citing why they like using computers was that computers could be used for writing stories. Another feature they liked were the printouts, which allowed them to immediately see a printed version of their stories. Some children also expressed their enjoyment of keyboarding or typing. There is no question that word processing can be used to interest children in both computing and written composition. Youngsters who use the microcomputer for word processing or graphics are motivated by the professional-looking products they are able to create. They report that they enjoy being able to make multiple copies of their stories or graphics to give to their friends and family. Students enjoy the software package Print Shop because of its communication capabilities. This ability to communicate easily with computers motivates all users, from elementary-school pupils to high-school students and adults.

There are many ways to motivate the use of word processing by making it part of educational activities that the students already enjoy. Word processing is a computer application that can integrate English language and communication skills with subjects such as the natural sciences, history, geography, and other social studies.

WORD PROCESSING AND EMERGING LITERACY

Word processing has a particularly positive effect on children just learning how to write. Young children are often discouraged because of the effort required to physically write the individual letters of the alphabet with their unsteady hands. Correcting errors by erasing paper-and-pencil work presents another difficulty to youngsters whose fine motor skills are not yet fully developed. Their word-processed draft forms are always neat and tidy, unlike handwritten drafts with their unsightly erasures and crossed-out words and sentences. The microcomputer allows the new writers to bypass their lack of hand–eye coordination and to immediately begin developing the skills of composition and written communication. Before computers, children's progress in writing was often hampered by the fact that they lacked the manual dexterity to actually write out the letters of the alphabet.

Word processing also helps youngsters to read because of the interdependence of reading and writing. "Software programs that contain text provided by programmers represent only one way in which computers can be used to foster reading with young children. Word processing programs offer many possibilities for building upon the connections between writing and reading by having students create and share their own materials. Simple data bases can be used in the primary grades to help students learn to search or gather and organize information. With music, drawing, and problem-solving software programs, children develop reading abilities as they see vocabulary, use commands, and follow directions to create songs and pictures and to solve puzzles" (Irwin 1987, 42). "We now have substantial evidence to indicate that there exists a dynamic relation between reading and writing, because each influences the other in the course of development, and that reading comprehension is engaged in during writing (through reading one's own writing) and is not a trivial matter" (Teale and Sulzby 1986). A child will naturally find it easier to read his or her own writing if it has been word processed rather than handwritten. "Children sounded out or pronounced letter names much more often when on computer, paused to reread more often, and generally revised their typing more often too" (Polin 1990a, 34–35).

The results of an annual and longitudinal study of children who had been through a Writing to Read program in kindergarten showed they outperformed a comparison group in reading, writing, and spelling (Polin, 1990a). Although word processing has been found to have a positive impact on emerging literacy, it is still used in only a small minority of cases. Only eighteen (or 14.5 percent) of 124 surveyed preschool programs in one county of Wisconsin used computers in their programs.

COLLABORATION, COMMUNICATION, AND SOCIAL INTERACTIONS

"Perhaps the most salient feature of writing process instruction is the collaboration that occurs among students and teachers within the writing community" (Montague 1990, 7). Observations at two schools (Sullivan et al. 1986) revealed that the role of the teacher was absolutely key to the success of word-processing activities in the classroom, particularly when students are first learning to use this tool. Word processing must also be successfully integrated into the students' program and presented in a way that appeals to their imagination.

The teacher must understand the nature of word processing and that it represents a new medium, unlike the schoolteacher who only accepted the use of word processing in her class with the following proviso: "Remember, once the students have finished word processing their stories with the computer, they will then have to copy their stories into their notebooks." Although this teacher agreed to use the new technology, it is obvious that she did not fully understand its implications, nor was she able to give up her traditional notions of primary education. Fortunately, this teacher's attitude is more the exception than the rule. The level and type of interaction between teachers and students seems to change when word processing is introduced, which seems to contribute to an overall improvement in writing (Burnet 1986). Even among pupils who were first learning to write, collaboration was an important ingredient in their learning experience. "One important source of literary activity for young children is collaboration, either with adults or peers, through sharing and watching. A lot of interactions took place at the computers, little of it true collaboration. Most talk involved a waiting student who might provide audience commentary or technical assistance. First graders developed and made use of a culture of computer expertise in which any student could be a resource. Whenever one child discovered a new function . . . the news spread quickly" (Polin 1990a, 34–35).

WORD PROCESSING AND STUDENTS' COMPOSITION SKILLS

Although research into the impact of word processing on development of literacy skills is still primitive, the overwhelming conclusion of almost all researchers is positive. "Many studies document the increased length of student texts when produced by word processors rather than by hand . . .

These studies replicate with digital technologies Ben Wood's findings on student writing with typewriters half a century ago . . . [Students who] learned to write on portable typewriters . . . when compared with a control group who wrote without typewriters . . . wrote more and with more expression, advanced in reading scores, became better spellers, and expressed greater interest in and enjoyment of writing" (Pea and Kurland 1986). In addition to students producing longer texts, many researchers have also reported an improvement in the quality of the writing produced with word processors (Bruce and Rubin 1984; Montague 1990; Moore 1987), although these results are sometimes mixed with greater improvements being posted by those who were stronger writers using traditional methods (Bryson et al. 1985).

One of the expectations of researchers was that word processing would increase the amount and quality of revisions made by students. This apparently was not always the case, as a number of studies reveal. Although students do not seem to make more revisions with a word processor than with a pen, they do revise differently (Daiute 1986, 141–59). They tend to add text to the end of compositions and to increase their fluency, measured as the total number of words produced. A similar phenomenon occurs with planning. There is no increase in the planning phase with word processing as was expected. The reason seems to be that students are willing to plunge into a composition without extensive preplanning because of the ease with which the text can be revised.

WORD PROCESSING AND THE LEARNING-DISABLED

Word processing with learning-disabled children is one of the more successful educational applications of computers.

> Children used either Storywriter or Paperclip, depending on their developmental levels and their writing facility . . . Toby [the teacher] felt that the children's writing had improved a great deal because they could get a neat finished product, and they enjoyed writing much more than previously. [She says] "The computer

> makes things easier, which is particularly important for these children because they have fine motor problems and have trouble organizing their thoughts." On one occasion, children's output on the computer was compared with their output while writing with pencil and paper, confirming the teacher's observation that they wrote more when using the word processor. Of the children in the class, all but one expressed a preference for writing on the computer. (Sullivan et al. 1986)

Our observations have been confirmed in a more extensive study comparing pencil and computer outputs of fourteen fourth-grade learning- disabled students (Morocco and Neuman, 243–47). As was the case at our site, the main reason for success was the teacher's ability to interact with the students as they produced their text. At our site, the teacher was able to interact with five or six students at a time by making frequent stops at their workstations and commenting on the text that showed on the monitors. The students obtained much more attention than they would have if they were writing with pencil and paper because of the difficulty the teacher would have encountered reading their texts.

The microcomputer opens new worlds of expression to students and helps them develop a positive attitude to writing and, consequently, to learning in general. Because of its ability to simplify writing and facilitate editing, word processing has had a positive effect on the writing skills of students of all ages, from the earliest grades of primary school to university graduate schools. Word processing appears to have the potential to arrest or even reverse the general decline in literacy and writing skills that McLuhan associated with television and its influence. The widespread use of microcomputers for word processing should result in a renewal of our literary roots and the restoration of the pretelevision level of reading.

ROBERT K. LOGAN

Science Education: Microcomputers, Spreadsheets, and Databases

The recognition of the importance of science education has been steadily increasing in recent years because of the vital dependence of our economy on technically and scientifically trained workers. Despite the increased need for excellence in science education, numerous studies show that contemporary schools do not provide adequate training in science and technology (Science Council of Canada 1984, 25, 33).

A MODEST PHILOSOPHY OF SCIENCE EDUCATION

The goals for science education should be modest and realistic. The primary objective of science education in elementary and secondary school should not be the dissemination of scientific knowledge, but the preparation of a child for a lifelong habit of scientific inquiry, using the processes and methodology of science. A second objective is to introduce the child to the phenomena of science and nature. A third is to stimulate within the child an enthusiasm and interest in science at both the phenomenological and methodological levels.

Science education, particularly in our primary and secondary schools, has been criticized for too much emphasis on content (or scientific facts) and too little emphasis on the experiential, experimental, and process dimensions of science. It is probably true that schoolchildren do not have enough opportunities to study nature firsthand and that their understanding of science and the scientific method is limited because they learn too many things secondhand instead of discovering scientific principles for themselves.

I see no conflict between the advocates of process-oriented science education and those who focus on content guidelines. Children must be introduced to both the processes of scientific inquiry and the content of scientific knowledge. In fact, each is absolutely necessary for appreciating the full value of the other. Learning scientific content without understanding the processes by which the information was gathered is an empty and meaningless exercise. Learning the processes of science without becoming acquainted with the phenomena scientists study is equally sterile. The computer, by virtue of its powers of organization, can help to reconcile the artificial conflict between those who focus exclusively on content or exclusively on process. Students who are taught how to use a computer to process scientific data or content can integrate the two approaches so that both are "learned simultaneously by having both arise out of school investigations and active problem solving activities" (Litchfield and Mattson 1989, 39).

As elementary-school science education can only be of an introductory nature, this conflict is relatively unimportant. The primary goal is to prepare the child for a lifetime of scientific (disciplined) inquiry. Perhaps the most important preparation for a life of scientific study is the acquisition of the basic skills of the 3 Rs. Literacy and numeracy are absolutely essential to scientific inquiry, both from the point of view of an individual's development and that of society as a whole.

Abstract science, deductive logic, and axiomatic geometry began in ancient Greece partly due to the influence of the phonetic alphabet (Logan 1986a). The place number system and the concept of zero played a major role in the development of modern science during the Renaissance. The cognitive tools that arose from the development of alphabetic literacy and place number numeracy led naturally to scientific thinking.

Just as the physical development of the human embryo passes through all the stages of human evolution (ontogeny recapitulates philogeny), the cognitive development of the child passes through the same stages of intellectual development that humankind evolved through historically. This would argue for the notion that a grounding in literacy and numeracy is essential for the eventual scientific training and development of a child. Education in elementary schools must focus on these basic skills, which are essential not only for science and scientific inquiry, but for the student's general education.

As children develop their basic cognitive skills by learning to decipher and exploit mathematical and literate media for information gathering and problem solving, they can be learning science at the same time. The use of those skills is essential for becoming acquainted with the basic phenomena of nature as well as scientific thought and culture. Combining science education with the acquisition of the basic skills of literacy and numeracy reinforces the prevalent philosophy of elementary education and forms one of the building blocks of the philosophy of science education I am advocating. Because electronic-information technologies provide a natural environment for the integration of knowledge (as opposed to print, which tends to fragment and specialize knowledge), the microcomputer is an ideal medium for an integrated approach to science education.

In addition to fundamental literacy and numeracy skills, the following process skills of science are essential elements of both a child's science and general education: classifying; seriating; measuring; inferring; predicting; interpreting; hypothesizing; and controlling variables. Disciplined inquiry is perhaps the most important lesson students can derive from their science education. Learning in the early years is facilitated through the manipulation of three-dimensional palpable objects rather than the formalistic study of the two-dimensional abstraction of words and numbers. Counting on one's fingers is a simple example; playing at a water table or molding clay is another. Science programs must, therefore, give the highest priority to exposing children to natural phenomena directly in the field or through a laboratory setting. The use of audiovisual media, including computer simulations, should be used in a supplementary manner to extend the breadth and depth of the child's exposure to scientific phenomena and is appropriate as long as it is not a substitute for the student's direct concrete experience of nature. Computers are most useful when employed by children as a tool to process the empirical data they have gathered experimentally on their own. The philosophy of science education which I have outlined for elementary-school science programs applies with equal validity to secondary- and postsecondary science education. No matter what the age of the student, learning the processes of science and being acquainted with the phenomena of nature are the surest way of understanding science and setting a foundation for a lifetime of scientific inquiry. As students mature and develop their cognitive skills, the level of sophistication of their scientific studies will increase, but they will still be concerned with the same basic skills of classifying, seriating, measuring, inferring, predicting, interpreting, hypothesizing, and controlling variables.

A PHILOSOPHY OF THE USE OF COMPUTERS IN SCIENCE EDUCATION

My philosophy of computer use in science education is based on the philosophy of science education presented in the previous section and consists of the following points.

1. The computer should never be used as a substitute for the student's hands-on contact with the phenomena of nature.

2. Having established that the computer should play a support role in science education, its foremost application will be in the tool mode, as the ideal medium for performing calculations, word processing, database management, spreadsheet analysis, and graphics. Databases increase the information available to students and provide them with a tool to organize information they have gathered on their own. Spreadsheets and graph-plotting routines provide valuable tools for processing and displaying scientific data.

3. Another support function that the computer can play will be to extend, through simulations, the breadth and depth of exposure to the scientific phenomena that students cannot manage on their own. For example, a simulation of the internal processes of a plant or an animal would be an appropriate supplement to a student's direct observation of the life form. Simulations can also artificially create flexible universes or microworlds with which the student may cybernetically interact and hence discover some of the properties of the real world. Computers can also be used to explore the real world through automated instrumentation which facilitates the student's gathering of empirical data.

4. The use of the computer for CAI, drill-and-practice, or tutorial should not play a prominent role if one wishes to emphasize the process rather than knowledge orientation of science education. Unfortunately, "the most commonly used programs in junior-high math classes and in

science classes at both junior and senior high levels were drill-and-practice programs for practicing algorithms, learning vocabulary and substantive factual information" (Becker 1991b, 20). Some uses of CAI can be envisioned, however, such as exposing the student to the basic vocabulary of science and some rudimentary facts.

5. The operations of the computer itself can become an object of investigation and disciplined inquiry that acquaints the student with the methods and processes of scientific study. The computer is also a perfect tool for demonstrating the power of technology. By coupling the computer to robotic devices (or even three-dimensional turtles in the case of LOGO), students can be exposed to mechanical devices and the concept of process control.

In conclusion, the computer should never be used as a substitute for the student's direct contact with or experience of the phenomena of nature. The computer should play a support role to a process- and phenomenon-based science program and never serve as the primary deliverer of science education. Where possible, students should use the computer the way a scientist does, namely, to organize and process raw data. Science is organized knowledge; thus the computer as a natural organizer of information can be used as a powerful tool for both science and science education.

COGNITIVE DEVELOPMENT THEORIES, COMPUTERS, AND SCIENCE EDUCATION

To ensure that computers play an appropriate role in science education, it is important that their use be guided by our understanding of the cognitive development of children's scientific abilities, as is exemplified by the approach of the Nuffield Junior Science Project:

> Children's practical problem solving is essentially a scientific way of working, so that the task in school is not one of teaching science to children, but rather of utilizing the children's own scientific way of working as a potent educational tool . . . Their own questions seem to be the most significant and to result most often in careful investigations. (Her Majesty's Inspectors of Schools 1984, 34)

Many of Piaget's findings (Piaget 1950) are worth keeping in mind when planning the use of computers to support the science education of children:

1. Children pass through a number of stages in their intellectual development.
2. While the time in life that these stages appear varies from child to child and the average time of their appearance differs from society to society, the order of the stages of development is a constant and a child cannot pass through any of the later stages until the earlier stages have been completed.
3. The stages are gradual and continuous and there are no sudden discontinuities in the child's development.
4. Children learn by acting on the world and transforming it and thereby coming to partial understandings which are continually revised, broadened, restructured, and related to each other.

Piaget's work reinforces the notion that science cannot be taught by cramming scientific facts into a child's head, nor can it work by merely teaching children the concept of the scientific method. "As far as education is concerned, the chief outcome of this theory of intellectual development is a plea that children be allowed to do their own learning" (Duckworth 1964, 172–75). Children must come to learn the structure of scientific thinking by processing information themselves. Unless they experience the processing of the data within their own psyche, there can be no cognitive development to the formal operational stage, which is the prerequisite for true scientific thinking and disciplined inquiry. Obtaining the "right answer" is absolutely useless unless the children participate in the process of discovering it for themselves.

In view of these considerations, the best use of computers in science education is for students to use them in the tool mode to organize and process scientific data they themselves have gathered or generated. The use of drill and practice should be limited to the acquisition of the basic vocabulary of scientific discourse. The memorization of scientific facts or processes is counterproductive as it reduces the time available for the direct observation of nature and the internal processing of information required for a meaningful science education.

Particular attention must be paid to the fact that young children require concrete experiences to absorb scientific concepts and yet the computer operates largely in the domain of the abstract. This is the reason that the computer should play a supplemental role only and must always be subordinated to processing data gathered from concrete observations or experiences. Wherever possible, the coupling of the computer or the integration of it with natural phenomena occurring in real time is always to be encouraged. Robots can play an important role, for example, because they provide a link between computer functions and movement in three-dimensional space in real time.

The computer is an ideal medium for providing youngsters with the opportunity to invent and discover on their own, whether it is the inner working of the computer itself or the information stored in the computer's memory. In order to make the experience successful, however, the child must play a role in the creation of the data. Otherwise, the child is merely processing abstract data unrelated to his or her own experience. Another aspect of the computer which makes it a useful tool for science education is the natural way it promotes the same collaborative spirit among its users as it does among science and engineering researchers. The computer is well known for its ability to promote peer learning and interactions among its young users, as shown in Chapter 5 of this book. The communication function of computers contributes another important ingredient for science education.

SCIENCE EDUCATION AND THE TOOL MODE

The tool aspects of the computer should be exploited by the child in order to gather, record, analyze, organize, and interpret scientific data and information. The challenge to a teacher, therefore, is to find specific applications of basic tools like spreadsheets and databases to science education.

1. Database Software

A database is any collection of information. The very first databases were created in the schools in Sumer some 5,000 years ago when teachers compiled on clay tablets lists of all the known trees, animals, rivers, mountains, and kings. These lists of items belonging to the same category were used for lessons and constitute history's first databases and first courseware. The management of the first databases and subsequent ones up to the advent of computers was performed manually by the user. With computers it is possible to store and collect large amounts of data and then to automatically search, sort, analyze, classify, reorganize, present, and re-present that information with a few basic menu-driven commands. One should really distinguish between the database management tools such as dBaseIV, Paradox, or AppleWorks and the actual database or collection of information. The term "database" is now used interchangeably to denote either the tool or the collection of data, and is most often used to denote the combination of the two.

Database programs can be applied in one of two modes, passively or interactively. In the passive mode, a prepackaged database can be employed as a source of information which the student draws upon as he or she studies a particular topic. In the interactive mode students use a database management program to create their own database of information, which they may or may not use for some other educational activity. The educational experience in this case is not limited to the content of the database, as is the case with the passive mode; rather, students learn new things through the process of organizing their own information. Teachers who have used databases were quite adamant about this point.

> Students should be given opportunities to be able to set up databases, draw inferences, etc. Data entry is not important. Database creation can generate very open ended exploratory types of activities. They enable students to gather, sort, and classify information and manipulate variables. They can develop higher order thinking skills in this manner. (Logan 1986b)

Before organizing one's own database, it is helpful to add new entries to an existing database using preexisting organizational structures. The next step is to design a database from scratch. Such an exercise is an open-ended exploratory activity that will help the student develop two sets of skills: the technical skills needed to use the database software, plus a number of higher-order thinking skills needed to analyze and interpret information including sorting, manipulating variables, classifying information, and drawing inferences.

A well-organized database program can help students see relationships among the data they have collected which they might not ordinarily have seen. The database can help the student discover commonalities, similarities, or differences between the elements of the database. They are in a better position to analyze the various relationships that might exist as well as note trends. They might even be in a position to formulate and test hypotheses. "The capability of organizing and rearranging information in a multitude of formats offers the user the opportunity to view the data in a variety of ways. Using databases to perform queries allows relevant bits of information to become visible, while leaving the database unchanged" (Strickland 1989, 20). In summary, the construction of a database is an ideal activity for the student to learn the skills of classifying, seriating, inferring, predicting, hypothesizing, and controlling variables.

Randy Saylor, a consultant with the Toronto school board, used the Acorn computer with a grade 6 class in order to classify animals. Randy motivated the pupils to use the database system by having them first classify things with which they were familiar and which were important to them, such as names of candy bars, rock-and-roll stars, and television programs. He then divided the class into research teams which created databases on mammals, breeds of cattle, breeds of cats, snakes, and other reptiles.

The use of the AppleWork database in conjunction with a high-school chemistry laboratory in Eastern Montana created an application in which students were able to "form and test hypotheses and actually 'do' science" (Strickland and Hoffer 1990–91, 30). The students first created a database of the physical properties of chemical compounds choosing those properties which would facilitate the identification of the compounds in the laboratory. They listed properties such as physical state, melting point, color, solubility, and density. Each student was then assigned three compounds that they were responsible for researching and adding to the database. In this manner, "an 'electronic chemical handbook' containing information about the physical properties of over 200 different chemical compounds was created" (Mandinach 1989, 20). The students were then assigned problems which they could answer by making use of the database and which required them to analyze, synthesize, and evaluate the information contained in the database.

Relational databases have also been used with equal effectiveness in the humanities. A teacher who used databases for a high-school English program reached the following conclusion:

> Databases are powerful tools. They can pose interesting questions without "right" answers and pose problems that have multiple solutions or no answer at all. Our society needs thinkers and writers who can work with these kinds of questions and problems. By using database projects as the stimulus for thinking, teachers can go a long way toward seeing that society's needs are met. (Blakenbaker 1987, 17–18)

In addition to the cognitive development it promotes and the assistance it provides to science education, learning to use a database program may have its own intrinsic value. "As we move towards the 21st century information processing may become a survival skill. Many researchers and educators believe that searching databases may well become one of the most important skills we can teach . . . Database searching, retrieval and design are tools for the instructor to provide motivation, to develop deductive reasoning and critical thinking, and to reinforce concepts and information within a particular discipline" (Mandinach 1989, 27).

2. Spreadsheet Software

One of the chief advantages of spreadsheets is that they allow users to visualize the information or data that they wish to understand. New relationships emerge once data is presented visually. Another advantage of the spreadsheet routine is that it allows users to easily make calculations involving the data entered into the spreadsheet. This capability facilitates inferring, predicting, and hypothesizing.

> Although the fear of mathematics is a documented problem among teachers as well as students, the time consumption and sheer tedium of organizing, manipulating and interpreting large amounts of numerical data make even math and science teachers avoid certain instructional experiences as being too "effort-costly" for the classroom. However, working with quantified information adds important dimensions to every content area, so the effective use of computers in making number processing faster and easier can facilitate student acquisition of subject matter and improve quantitative thinking skills across the curriculum. Equally important, though, is the fact that the number-processing power of computers can also allow students conveniently to experiment with mathematical models of relationships among given variables by changing specific quantities repeatedly and noting the effects upon related variables . . . Students develop new technology-centered modes of approaching and solving problems, and the computer becomes a tool with the potential to truly transform the curriculum. (Dyrli 1986, 47)

The spreadsheet is a useful tool because it places the abstract data collected by the student within a more concrete context, namely, the graphical cells of the spreadsheet. In order to make the spreadsheet concept itself more concrete, students should be encouraged to create visual arrays of data or information on paper.

When I first began my study of spreadsheets, their use in classrooms was extremely limited. In order to gauge the practicality of using this application in elementary schools, I organized the use of a spreadsheet by a grade 5/6 class at the Kensington Community School in downtown Toronto in a lesson which lasted an hour and a half. The day before the spreadsheet lesson, the teacher, Olga Panowic, had the children create a chart with the name, age, height (in centimeters), and weight (in newtons) of each child in the class. The session began by having the children call up the spreadsheet program on the computer. I then introduced them to the concept of a cell and taught them to move from cell to cell. We then played "spreadsheet bingo" in which cell numbers (C5, J28, AB21) were called out and the children moved their cursors to the cell, called out, and raised their hands.

The children then learned how to create labels and entered those that were needed for our exercise. They began by entering their names, ages, heights, and weights. They then converted their weights in newtons to their weights in kilograms and pounds using the calculating capacities of the spreadsheet software.

After these preliminaries were completed, we held a broad-jump contest and the pupils learned how to measure the distance they jumped. They then entered the distance they broad-jumped on the spreadsheet, after which they calculated the ratio of the distance jumped divided by the height of the jumpers. One of the taller boys who was bragging how far he'd jumped was taken down a peg when he learned that unlike many of the others, his ratio was less than 1.0, which meant that he was unable to jump a distance equal to his height.

In the hour-and-a-half lesson, we were able to help the students learn how to measure distances, convert from one set of units to another, and calculate and understand the meaning of ratios. One of the reasons for our success was that the data we collected and analyzed were so graphically displayed by the spreadsheet. The field trial was a success in that the pupils learned many things in a short time. Spreadsheet routines are amenable to a number of applications pertinent to science education. Other applications can be easily developed wherever there is a need to work with numerical data.

> Even at the most basic level, a spread sheet can be set up to organize any numeric information in a subject area and perform necessary calculations: story problems in mathematics; chemical

> reaction problems in science; state and national income/expenditures in social studies; word-frequency counts in language arts; balance sheets in business; track and field statistics in physical education; popular record sales in music; and household budgeting in home economics education. A good place for a teacher to begin in experimenting with instructional applications, therefore, is to go through textbooks and curriculum guides with a view to selecting and adapting quantified information for spreadsheet presentation, manipulation, and analysis . . . In addition to giving students experience with quantified data per se, spread sheets offer learners another tool for playing with ideas, and they underscore the importance of numerical data as a special dimension for knowing a subject and communicating about it. (Polin 1991, 9)

The strength of spreadsheets is their ability to display a "visual representation of data in tabular form." Spreadsheets also have the capacity to promote the skills of classifying, seriating, inferring, predicting, and controlling variables. These capabilities are particularly valuable in that they provide both a qualitative and quantitative dimension to the analysis and display of data. While spreadsheets and other software packages are powerful tools for learning mathematics, there is still a role to be played by pencil and paper, as the following results from a survey of elementary-school pupils indicates. When asked if it was easier to do math with computers or pencil and paper, the children were divided. Of the twenty-three responding to the question, fourteen preferred computers, and nine pencil and paper. The children who chose paper and pencil had nothing negative to say about computers, but expressed the sentiment that they needed the paper for doing calculations. One youngster claimed it was "easier at the desk because I can figure it out on a piece of paper." Another said, "It's better to do math with a pencil and paper, because your hand is right by the number, and you can make better sense of it." The computer does not permit doodling and this is an advantage inherent to working on paper. A large number of children offered no opinion because they had not done math extensively on computers. A similar attitude was found among secondary-school students. "Handwriting continued to be the first choice for many students for a number of reasons. Students found the laptops unsuitable for math and science classes because of the difficulty creating tabled data, diagrams, or math equations" (Polin 1990c, 7).

3. Word Processing

In addition to discovering scientific principles from the firsthand observation of nature and by processing the data that has been collected, students should also engage in communicating their results, evaluating and summarizing the results of others, and applying scientific knowledge to practical everyday problems. One of the computer-based tools which can assist them in these activities is word processing. By combining word processing and graphics with other tool packages, the student can synthesize and record information in the format of a science report which will give greater impact to their ideas. Word processing not only helps pupils communicate their thoughts about science, it also helps them organize information and think about new concepts. Word processing also facilitates the sharing of information. Just as the printing press played an important role in the development of modern science (Eisenstein 1979), so will word processing make an important contribution to contemporary science.

By combining word processing with spreadsheets, database management, and graphics, one could create a new software tool which I will call the "science report processor." Essentially, this utility would be a shell program based on word processing which would guide the student in the preparation of a science report. It could also have prompts reminding students that scientific reports should contain a description of the basic elements of science, namely, objectives, observation, measurement, data collection, inference, prediction, hypothesis, and methodology. The utility would help the student organize and classify data, perform calculations, estimate errors, make inferences, formulate and test hypotheses, and communicate results. The science report processor does not yet exist as a package; it is merely a concept for software which any of my readers is invited to develop.

An example of how word processing was used in a science project is provided by a grade 4 class in Virginia that was interested in conducting a scientific study of life-forms in a nearby pond. Their teacher used the opportunity to teach them some of the literacy strategies needed to conduct a scientific inquiry. They had lessons on developing strategies to find out information, to record what was learned, and to share their knowledge. They discussed what they would need to know to collect specimens and went to the library to do some research before going to the pond. "Notetaking strategies were also seen as purposeful by students who welcomed a means to record descriptions and identities of the specimens they found in order to facilitate their further study in the classroom. The computer played a most significant role in helping students organize and draft information in a form to be shared with others. Students used computers to assist them through all the steps in the writing process" (Feeley et al. 1987, 6).

THE COMPUTER IS THE MEDIUM AND SCIENCE IS THE MESSAGE

The medium of the microcomputer intrinsically promotes scientific methodology because of the logical way it operates. Word processing, databases, spreadsheets, and programming are examples of computer-based activities which subliminally incorporate the organizational patterns of scientific thinking. One might say that the computer is the medium and science methodology is the message.

It is not only the logic of the computer which parallels science, but the fact that it is a tool for self-learning and self-discovery. The methodology of science is nothing more than an organizational procedure for self-discovery and self-learning. The book of nature is not a volume from which scientists can merely read out the right answers. The laws of nature are uncovered through a tedious and careful process of observation and interaction with nature. This is similar to the process of using a computer in the exploration-and-discovery mode. To a certain degree, one examines the microworld of a software package the way a scientist explores the nature of the real world.

The computer, unlike the textbook, offers students the opportunity of creating their own curriculum. If students are provided with a rich enough computer-based learning environment, they will be able to make discoveries of their own as their interests move from one theme to another. The role of the science teacher will become that of an *agent provocateur* of the student's imagination by introducing the student to hands-on experiences with the phenomena of nature and the appropriate computer-based activities to complement these experiences. As we discovered from Piaget, making discoveries on one's own is the way children best learn science. Given these possibilities, coordination of hands-on, phenomenon-based learning and computer-based learning will be essential if the full potential of the positive change that computers can have on science education is to be realized.

Computers and Hybrid Systems: Networks, Multimedia, and Hypertext

Electronic systems, especially computers, lend themselves to hybridization. A hybrid system is one in which a mixture of media and/or technological devices are linked together to create a new application. A simple example of a hybrid system is the way in which a simple network can be created between two computers at remote sites by connecting them with two modems and the telephone system. Computers and their software, as was noted in Chapter 2, are themselves hybrid systems because of the way they combine into one system: a cybernetic central processing unit, magnetic disk drives for storing and accessing information, a video monitor to instantly display information, an electronic keyboard for inputting textual and numerical data, a scanner or koala pad for inputting graphic information, and a printer to output information on paper. As computers can be combined with almost any device requiring some element of control, ranging from automobiles to refrigerators, there is an almost unlimited number of hybrid systems to consider. This study, however, will limit itself to examining those hybrid systems which will have the greatest impact on education. These are:

1. networks including local area networks (LANs) and wide area networks (WANs),
2. multimedia systems including CD-ROM and videodisc technology, and
3. hypertext and hypermedia.

Hybrid systems are already having a significant impact on the classroom, as the following reports indicate:

> Over the two years of Apple Classroom of Tomorrow students demonstrated growth in the sophistication of their approaches to writing. For example, students reported that they were more inclined to review and reflect on their computer based writings than on handwritten text. They also reported a greater willingness to share their written work with peers, and recognized the value of doing so to improve their work. In addition to the increased refinement and maturity in writing skills, students were more likely to explore topics and record ideas more fully. The most interesting findings are related to the increased use of the computer to generate alternatives to text-based representations of ideas and the increasing use of graphics with text documents. In the second year students reported the addition of hypermedia, interactive video and robotics technology radically changed the way they learned . . . Students were more actively involved in designing and building curriculum projects, and hence, more responsible for constructing their own knowledge. (Polin 1991, 9)

NETWORKS

In considering the use of computer networks, one must ask what advantages they have over conventional face-to-face communication. One obvious advantage is that they allow children whose interactions are almost exclusively local to interact with children from other communities and even distant lands.

Not only do networks allow communication over great distances, they provide control over time, as well. Network users are able to communicate with each other at convenient times without concern for arranging a rendezvous or common meeting point, as is the case with face-to-face or even telephone conversations. Since computer users regularly return to their workstations, they are able to maintain a sustained dialogue with their network partners without being concerned about scheduling. Just as the water well or marketplace of a village became a convenient node where regular information exchange and social interactions could take place, so too can networked computers bring people together today.

The principal use of networks in schools is to distribute software to individual workstations by means of a file server. Unfortunately, only in a minority of cases are networks used to foster collaborative work between students. If a network is planned to foster collaboration, thought must be given to the organization and structure of the group, structure of the task, response opportunities, response obligations, and coordination and evaluation (Riel 1991–92, 5–7, 52).

A study of successful networks reveals that:

1. The less prior knowledge of an organization among the participants, the more important it is for the group to have a well specified group task.
2. Network participants need convenient reliable access to the network and/or external motivation for using the system (low cost, job requirement, connection to educational resources).
3. Networks need a leader, someone who is responsible for monitoring and facilitating the group interaction. (Riel 1991–92, 5–7, 52)

An example of a networking software that fosters collaboration is the Earth Lab Project developed by the Bank Street College as a local area network (LAN) which facilitates communications among students and their teachers. One of the reasons that the Earth Lab increased group work is that "the network interface carried some of the organizational burden of group work by providing common templates for groups to use in gathering and preparing data for analysis and for sharing data with others" (Polin 1990b, 5–6). A second factor that contributed to Earth Lab's success was that "the network gave teachers access to the existing organization of student groups, and facilitated the distribution and monitoring of group tasks." Earth Lab allowed students to investigate a number of interesting topics and relationships within the context of earth science, including weather patterns, location of dinosaur fossils, plate tectonics, volcanoes, and earthquakes. Students gathered data relating to the different aspects of a problem and then combined their results. Earth Lab provided the framework "for communications that elementary students could use as members of a science community comprised of peers and teachers . . . In the 'real' science community, science is a social activity in which scientists share data, hunches, findings, and theories with other scientists, or work collaboratively on a team" (Polin 1990b, 5–6). Earth Lab recreates this kind of environment, allowing students to develop an understanding of how science and scientists work.

An international computer network, the InterCultural Learning Network, linking children in the United States, Japan, Mexico, and Israel, has been used to develop collaborative science projects (Levin et al. 1987, 254–60). Activities were organized around projects in which a common problem was analyzed. Comparisons between different sites were made to obtain a more global perspective on local problems. The Water Problem Solving Project tackled the shortage of drinking water. Students in different parts of the world compared ways in which their water supplies were obtained and what techniques were used to conserve them. Techniques and sources were compared and analyzed to determine if they could be transferred from one site to another. The Stormy Weather Project studied severe weather conditions and techniques to prepare for and survive these natural catastrophes. The project What a Pest compared local pests and the techniques for combatting them. The Pollution and You Project cataloged local air and water pollution, toxic and other waste disposal challenges, and techniques for dealing with these problems. Comparisons were made of the procedures used to handle these problems to determine how local procedures could be improved.

These projects provided students with a way of carrying out the activities of professional scientists, involving a complete range of science skills including data collection, data analysis, pattern recognition, hypothesis generation and testing, collaboration, and communication.

Another example of international collaboration is the Computer Chronicle, a computer-networked interschool news service which links children from California, Alaska, Hawaii, Mexico, and Israel (Mehan 1989, 4–22). Students composed stories which were placed on the network and eventually downloaded, edited, and placed in school newspapers at one another's sites. One of the advantages of the system, in addition to enhancing cross-cultural communications, was that the children were composing stories for other children and not just creating academic texts for their teachers. The fact that real peer communication was taking place motivated the children to develop their writing skills. Standardized achievement tests revealed that students who participated in the Computer Chronicle Project gained two to three grade levels in their language-arts skills in one year.

Similar results were obtained in the InterCultural Learning Network project with fourth-graders, as verified by school-administered, standardized achievement tests. In the year of the project, the class went from slightly below their grade level in the language area to almost one full year above their grade level. In a study with seventh-graders, two compositions were compared, one written for peers on the network and one written for the teacher. The composition written for their peers received a "significantly higher rating." "The development of functional writing environments to contextualize students' work can lead to significant improvements in the quality of students' classroom writing" (Riel 1991–92, 5–7). Another advantage of networks in classrooms in which pupils and the teacher are on-line is that there is less competition for the teacher's attention.

The hybridization of the microcomputer with other presentation media including text, music, audio, and video has created a new educational medium, multimedia, which is finding its way into the classroom. One example is the Bartlett Family, a multimedia learning package created for the Ontario Ministry of Education, which combines a standard textbook and dramatized video presentations of episodes from pioneer life in Ontario with a microcomputer-based software program.

New hardware and software systems are now on the market to address the needs of those who wish to create multimedia presentations. Among the new products which have recently surfaced are Macromedia Director, Adobe Premier, and Commodore's Amiga 3000 series microcomputers. The Macintosh program HyperCard, which has database features built into it, has given rise to another specialized form of multimedia known as hypertext and hyper-media, which will be discussed later.

One of the advantages of a multimedia educational package is that the author may select the appropriate medium for each educational objective he or she wishes to achieve. There is also the advantage that information conveyed in one medium is reinforced when conveyed in another medium. Multimedia presentations help students for whom reading may pose a barrier to learning; by experiencing the same information in both a video format and in text, these students may overcome some of their difficulties with basic literacy.

A key factor for the success of computers in education has been their ability to compete successfully with television for the attention of young people. Multimedia applications which combine software-based learning tools with text, graphics, and video enhance the learning package's visual appeal and are in an even stronger position to compete with commercial television than a computer software package by itself. Unlike television, multimedia packages incorporating video are interactive and provide a viable learning environment. Because textual material is integrated into a multimedia learning package, they encourage more children to read. Multimedia packages, like computer software, are also an excellent medium for individualized learning as students can proceed through the material at their own pace and choose their own paths depending on their interests.

Multimedia learning environments, also like traditional computer software, seem to motivate a general interest in school. At New York City's School of the Future, where 150 seventh- and eighth-graders study in classrooms which make heavy use of multimedia learning environments, the attendance rate is 98 percent, compared with a citywide average of 86.5 percent (E. Schwartz 1991, 158). Multimedia learning packages encourage strong social interactions because of the ease with which information can be shared. They are also excellent tools for integrating the curriculum as the discussion below of the Voyage of the Mimi, a multimedia learning package, reveals. In short, multimedia learning packages share all of the positive features of traditional software, and hence enhance the way in which computers are reforming education.

The most powerful factor in multimedia's reformation of education is the control students have over the flow of information in content and format. Students are therefore able to find that unique path through the information which makes the greatest sense to them. One of the shortcomings of traditional education was that almost all of the learning had to be channeled through the narrow passage of literacy, which for some students was a major stumbling block or, in extreme cases, a total barrier. Computers have been revolutionary because they provided the first real alternative to print which encouraged learning and permitted cognitive development. With multimedia learning environments, the number of passages to learning is even greater because of the many choices it gives the student. There is the additional benefit that ideas, concepts, and information presented in one medium are reinforced in another. Just as a public event in which we have participated seems more real if we see it on television or read about it in the newspapers, so do ideas encountered in class seem more real when experienced in a variety of media.

The greatest contribution to the reform of education with multimedia will not arise from students using the medium to encounter and absorb new ideas and information, but rather from using the tool to create their own presentations. One of the great educational benefits of traditional computers is that students can create their own professional texts and thereby break down the monopolies of publishers. As Leibniz once remarked, the freedom of the press belongs to those who own one. Today, owning a computer and a laser printer is equivalent to owning a printing press, which is why nondemocratic or totalitarian states have controlled their use.

Multimedia software, of which HyperCard is the most easily accessed, allows students to create their own multimedia presentations. It even provides them with the power of a commercial television station if they can find a computer network over which they can distribute their video creations. This capability of multimedia applications not only will contribute to the reform of education, it might also demystify television for children and hence loosen its grip over them.

The organization of a multimedia learning package is very important for its success as an educational tool. If the information is presented in a linear fashion in which the student is guided through the material in a predetermined manner, the power of the medium is lost; the multimedia package basically degenerates into a textbook in which the graphics supplementing the text include audio and video material and the questions and problems at the end of the chapter have been replaced by a CAI, drill-and-practice, or tutorial format. If, on the other hand, students are able to interact with the learning material and participate in decision-making strategies and inquiry, they are more likely to develop higher-order intellectual skills and reasoning capabilities (Litchfield and Mattson 1989, 37).

An example of a multimedia learning package that allows students to explore on their own is the curriculum unit, the Voyage of the Mimi, developed in 1985. The components of the Mimi consist of the television series "*The Voyage of the Mimi* (thirteen dramatic episodes and thirteen expeditions), *The Voyage of the Mimi* book, wall charts, and four learning modules: Introduction to Computing, Maps and Navigation, Whales and Their Environment (with the Bank Street Laboratory), and Ecosystems. Each module contains teachers' guides, students' guides, and software" (Manzelli 1986, 55).

The richness of the Mimi multimedia package allows for the integration of almost the entire curriculum of study for middle-school students. One teacher described how she used the theme of the Mimi and some of the materials within the package to create a learning experience for the students. Because the videos document the way in which scientists gather information and collaborate, each member of the class was asked to choose one of the characters and keep a log of their activities in all of the episodes. The students then discussed with one another the nature of their characters and the feelings they imagined their characters to possess.

The students learned and absorbed many ideas in science by carefully watching the videos and doing follow-up research to understand what was observed in greater detail. The students learned a great deal of math from the section on maps and navigation, and developed their problem-solving skills in the context of topics that absorbed their interest and attention. They practiced math and problem-solving skills, for instance, by estimating the whale population in the waters off Boston. They then extended the techniques they learned to estimate Boston's squirrel population. The students also explored weather forecasting and kept track of local weather patterns. They went on field trips related to whales and the whaling industry. They studied the behavior of whales and so learned some zoology. They even integrated their social studies into the unit by studying the history of whaling and the lifestyles associated with it from the seventeenth century to the present. Finally, they read literature relating to whaling and the sea, such as *Moby Dick* and *Treasure Island*.

The Mimi material was used in conjunction with the Bank Street Laboratory, which permits students to

> enter data and compare or contrast them with other data. The classroom becomes a science-exploring world which opens up a process for the way students operate. They can begin to ask the "what if" questions, identify variables, describe how variables are controlled and interrelated, and formulate hypotheses. Students can save their investigations on a disk and go back to look at them throughout the year . . . The computer can help move students from the concrete to the abstract. The Bank Street Lab sensors allow them to be curious about science, to conduct investigations, to be methodical, and to inquire and investigate interesting ques-

> tions. The teacher can use this environment to place students in small groups to conduct investigations. Each student has a task, and tasks can be rotated every class session. Everyone has the chance to experience being the scientist by either using the sensors, or by recording observations and suggestions as to how to conduct the experiment the next time. (Manzelli 1986, 58–59)

The success of the program was the way the teacher put all the pieces of the curriculum together; as she put it, "It is up to the teacher to develop a plan to use these materials [the Mimi program] to add the depth and dimension needed for students to learn" (Manzelli 1986, 56–57).

HYPERTEXT AND HYPERMEDIA

Hypertext and hypermedia connect textual and multimedia information, respectively, within a relational database so that there are multiple ways of accessing and linking the data. A hypertext system is a database in which information has been organized into smaller units of information called nodes which can be easily joined to other information nodes within the database. Hypermedia is an extension of the hypertext concept in which the nodes of information are in multimedia formats.

Because hypermedia includes hypertext as a special case when the only hypermedium is text, I will use the term "hypermedia" from now on to refer generically to both hypertext and hypermedia and I will refer to it as a medium even though it is an amalgam of many media.

Early reports on the use of hypermedia indicate that this medium will have an important impact on education and the use of computers in schools. The reason that hypermedia is potentially so powerful is that "information is organized by its relations to other information and not by discrete categories such as alphabetical order. The power of hypermedia applications resides in the links or relationships that students weave as they explore information" (Blanchard 1989, 23). Without associative links, the information that students encounter can become decontextualized, and hence the learning process is less effective or, in some cases, completely stymied. With a hypermedia presentation, however, each student is able to create his or her own pathway through the information assembled for them. The students are able to process the information within their own psyches and so discover it on their own terms and take ownership of it. This is not possible with traditional media in which the information is either decontextualized, and thus rendered senseless, or the author imposes his or her own organizational structures on the material. That structure may suit the author's way of thinking but not the reader's.

As the author of this text, I am guilty of such an imposition. But you, as my reader, are not obliged to read my text in a straight line from beginning to end. You might very well get more out of it by reading things in an order that makes sense to you or corresponds to your own particular interests. This was the style of reading advocated by McLuhan and suggested in his aphorism, "The user is the content." Bill Gates, one of the developers of hypermedia, framed this idea in the following manner: "People don't think and learn in a straight line, from one fact to another; they go off on a million tangents, because they're interested in a million things" (Gates 1988, 14–15).

Hypermedia lends itself to some unique applications which cannot be readily achieved with traditional software. It is being used as a tool to teach children and adults how to read. The reader becomes more involved with the text because of the ability to select different outcomes. In some applications, activities actually help the user learn how to read. In Alphabet Blocks, an animated elf sounds out each letter using digitized speech and at the same time is shown making the the correct lip movements needed to create the sound of each letter. With a click of the mouse, all the letters can be changed from uppercase to lowercase or vice versa. This application is particularly effective because of the way it combines visual and auditory information. An adult literacy program developed by Drexel University includes hypermedia for alphabet sounds and recognition, vowel use, cloze activities, and Storyline, which develops analytic reasoning based on the readings of stories.

A number of hyper-applications, developed for different learning activities in the sciences, social sciences, literature, and foreign-language instruction, combine audio, video, graphical, and textual information. *Grolier Academic American Encyclopedia* comes in a hypermedia version which takes advantage of optical technology for large-scale textual and graphical information storage. "Students can move quickly along their own research paths in the 30,690 entries of the encyclopedia and analyze as many as 10 articles on a single screen" (Blanchard 1989, 27).

The use of hypermedia for writing offers tremendous possibilities for original expression and comment on existing material. An author is now able to create multiple links between his or her material and explore relationships that were not possible with a linear text. Before hypertext, the only way of creating links was with the table of contents, the index, footnotes, the bibliography, and illustrations. The Intermedia program developed at Brown University allows a literary critic to work with original texts, annotate them, create links between them, and make reference to other authors, critics, biographical data, or historical information. Intermedia will encourage the retrieval of some of the literary genres that were more popular before the advent of the printing press — for example, compilations which are collections of the writings of others to which the commentator adds his or her own comments (Logan 1986a, 211–12). There will also be a tendency to produce more books like this one and those that appeared before the advent of printing in which the author composes the principal part of the text but adds the text of other authors for the purpose of confirmation.

Of all the uses of hypermedia, the most exciting and promising one I have encountered is the work of B. Stebbins, a physics and biology teacher at West High in Columbus, Ohio (Stebbins 1990, 7–73). He found that HyperCard was one of the few software products that adapted to his needs as a teacher and allowed him to build an application from scratch with relative ease. What made his particular application of hypermedia so innovative and successful was that he emphasized the students' use of the application to create learning modules and reports that could be used by other students to learn the curriculum material being covered in his classes. "By putting my students in charge of their biology class, they probably learned more biology than if I had been in charge" (Stebbins 1990, 35). By taking this approach, he achieved the goals he set out for himself as a teacher. "The teacher's job is to teach the students how to learn, encourage them to investigate different ideas, and then get out of their way by putting the responsibility for learning into their hands" (Stebbins 1990, 39).

Stebbins described how and why he achieved his goals: "Most teachers realize that they actually had to teach in order to fully learn many things for themselves. Using the idea that the best way to learn is to teach someone else, we have often put our students in charge of teaching various sections of the text in the various content areas. This caused those students who had worked on a specific area to become the experts in that area" (Stebbins 1990, 12–13). It also created a community of learners who could help one another. These were not the only effects of this approach. Students began to suggest to their teachers how assignments should be done. They explored new intellectual areas their teachers had not even considered. They built on one another's experiences and continued to get better at their work.

Hypermedia required the students "to think in logical steps and had the ability to grow as the user demanded more from it" (Stebbins 1990, 7). Stebbins concluded, "My students were able to learn anything they needed to know. HyperCard became the tool that allows students to organize material in a manner that makes sense to them" (Stebbins 1990, 13). It is "the tool that not only changes the way students use the classroom computer, but also dramatically changes how students learn" (Stebbins 1990, 8). The students became aware of their learning process because they "had to show each individual step of the process. In a sense, we caused them to slow down the thinking process so that each and every detail was thought out completely" (Stebbins 1990, 22).

Hypermedia is also an ideal tool for individualized instruction because students can make up missed lessons by going through a stack on their own. They can "look at the material over and over again, and more importantly, the student controls the pace of the lesson. Again, the student controls the learning, not the teacher" (Stebbins 1990, 43-47).

REFERENCES

Albright, W.F. 1957. *From the Stone Age to Christianity*. 2d ed. Garden City, N.Y.: Doubleday Anchor Books.

Anderson, John R. 1980. *Cognitive Psychology and Its Implications*. San Francisco: W.H. Freeman.

Anderson, M.A. 1990. Technology Integration for Mainstreamed Students. *The Computing Teacher* 18{4}.

Ashby, W.R. 1957. *An Introduction to Cybernetics*. 2d ed. New York: J. Wiley.

Aston, M. 1968. *The Fifteenth Century: The Prospect of Europe*. London: Thames and Hudson.

Bain, G., and R. Price. 1972. Who Is a White Collar Employee? *B.J. Industrial Relations*. Vol.10.

Bandle, O., H. Klingenberg, and H. Mauer, eds. 1958. Kategorien. *Lehre vom Satz*. Hamburg.

Bank Street College Project in Science and Mathematics Education. 1985. *The Voyage of the Mimi*. New York: Holt, Rinehart & Winston.

Barker, Thomas. 1987. Studies in Word Processing and Writing. *Computers in the Schools* 4{1}.

Baron, N. 1983. *Speech, Writing and Sign*. Bloomington: Indiana Univ. Press.

Becker, Henry Jay. 1991a. When Powerful Tools Meet Conventional Beliefs and Institutional Constraints. *The Computing Teacher* 18{8}.

_____. 1991b. Mathematics and Science Uses of Computers in American Schools. *Journal of Computers in Mathematics and Science Teaching* 10{4}.

Bender, B. 1975. *Farming in Pre-History*. London: J. Baker.

Bitter, G., and R. Camuse. 1984. *Using a Microcomputer in the Classroom.* Reston, Va.: Reston Pub. Co.

Blakenbaker, R. 1987. Databases in the English Class: A Valuable Lesson. *The Computing Teacher* 15{2}.

Blanchard, J. 1989. Hypermedia: Hypertext Implications for Reading Education. *Computers in the Schools* 6{3/4}.

Bledstein, B.J. 1976. *The Culture of Professionalism, the Middle Class and the Development of Higher Education in America*. New York: Norton.

Bloomfield, L. 1933. *Language*. New York: Holt.

Boissenaide, P. 1927. *Life and Work in Medieval Europe*. New York: Knopf.

Bonnington, Paul. 1995. *Internet World*. [April].

Bosaquet, E.F. 1930. *English 17th Century Almanacks*. The Library 4th ser. x.

Boulding, Kenneth. 1961. *The Image*. Ann Arbor: Univ. of Michigan Press.

Bowen, James. 1972. *A History of Western Education*. Vol. 2. London: Methuen.

Brougham (Lord). Speeches on Social and Political Subjects [1857]. In *History of Education: Socrates to Montessori* by L. Cole. New York: Holt, Rinehart & Winston.

Brown, J.S. 1985. Process versus Product: A Perspective on Tools for Communal and Informal Learning Environments. *Journal of Educational Computing Research* 1{2}.

Bruce, B., and A. Rubin. 1984. The Utilization of Technology in the Development of Basic Skills Instruction: Written Communications. Final Report Vol.1. Tech. Report No. 5766. Bolt, Beranek and Newman.

Bryson, M., P.H. Lindsay, E. Joram, and E. Woodruff. 1985. *Augmented Word Processing: The Influence of Task Characteristics and Mode of Production on Writer's Cognition*. Toronto: Ont. Inst. for Studies in Education.

Burnet, J. 1986. Word Processing as a Tool of an Elementary School Student. Ph.D. diss., Univ. of Maryland. *Dissertation Abstracts International* 47, 1183A.

Carbonnier, G. 1969. *Conversations with Claude Levi-Strauss*. London: J. Cape.

Coburn, P., et al. 1982. *Practical Guide to Computers in Education*. Reading, Mass.: Addison-Wesley.

Cole, L. 1950. *History of Education: Socrates to Montessori*. New York: Holt, Rinehart & Winston.

Cole, Sonia. 1970. *The Neolithic Revolution*. London: Trustees of the British Museum.

Coleman, Janet. 1981. *Medieval Readers and Writers*. New York: Hutchinson.

Coulmas, F., and K. Erlich, eds. 1983. *Writing in Focus*. New York: Mouton.

Cronin, Mary. 1994. *Doing Business on the Internet*. New York: Van Nostrand Reinhold.

Czernis, L., and R.K. Logan. 1982. Telidon as Pure Advertising — Implications for Attention Span. Paper presented at Social Impact of Telidon and Other Electronic Technology Workshop, 1-2 March, at Univ. of Toronto, and sponsored by Dept. of Communication, Ottawa, and the Culture and Tech. Seminar, Univ. of Toronto.

Dahrendorf, R. 1959. *Class and Class Conflict in Industrial Society*. London: Routledge & K. Paul.

Daiute, Collette. 1986. Physical and Cognitive Factors in Revising: Insights from Studies with Computers. *Research in the Teaching of English* 20.

Danzig, T. 1968. *Number: The Language of Science*. New York: Greenwood.

Dawkins, Richard. 1976. *The Selfish Gene*. New York: Oxford Univ. Press.

Day Lewis, Cecil. 1947. *The Poetic Image*. London: J. Cape.

Demsky, Aaron. 1972. Scroll. Jerusalem: *Encyclopedia Judaica*.

Dickens, A.G. 1966. *Reformation and Society in 16th Century Europe*. London: Thames and Hudson.

Drazin, N. 1940. *History of Jewish Education 515 BCE to 220 BCE*. Reprint 1979. New York: Arno Press.

Drucker, Peter. 1993. *Post-Capitalist Society*. New York: HarperBusiness.

Duckworth, E. 1964. Piaget Rediscovered. *The Journal of Research in Science Teaching* 2{3}.

Dyrli, O.E. 1986. Electronic Spreadsheets in the Curriculum. *Computers in the Schools* 3{1}.

Edmonson, Munro. 1971. *Lore: An Introduction to the Science of Folklore and Literature*. New York: Holt, Rinehart & Winston.

Eisenstein, E.L. 1979. *The Printing Press as an Agent of Change*. London: Cambridge Univ. Press.

Feeley, J.T., D.S. Strickland, and S.B. Wepner. 1987. Computer as Tool: Classroom Applications for Language Arts. *Computers in the Schools* 4{1}.

Fish, M.C., and S.C. Feldman. 1990. Learning and Teaching in Microcomputer Classrooms: Reconsidering Assumptions. *Computers in the Schools* 7{3}.

Fisher, Ernst. 1963. *The Necessity of Art*. Baltimore: Penguin Books.

Fribers, J. 1984. Numbers and Measures in the Earliest Written Records. *Scientific American* Vol. 250. No. 2.

Garrard, J., et al. 1978. *The Middle Class in Politics*. Farnborough, Hents: Saxon House.

Gates, Bill. 1988. Computers in Schools, Today and Tomorrow. *T.H.E. Journal* 16{12}.

Giddens, A. 1981. *The Class Structure of the Advanced Societies*. London: Hutchinson.

Gilman, Daniel. 1906. The Dawn of a University. In *The Launching of a University and Other Papers: A Sheaf of Remembrances*. New York: Dodd, Mead.

Glotz, G. 1926. *Ancient Greece at Work*. New York: Knopf.

Graff, H.J. 1979. *The Literacy Myth*. New York: Academic Press.

Gretton, R.H. 1919. *The English Middle Class*. London: G. Bell and Sons.

Gudschinsky, Sarah. 1968. The Relationship of Language and Linguistics to Reading. In *Literacy: The Growing Influence of Linguistics* by S. Gudschinsky [1976]. The Hague: Mouton.

Hall, Edward. 1969. *The Hidden Dimension*. Garden City, N.Y.: Anchor Books.

________. 1973. *The Silent Language*. Garden City, N.Y.: Anchor Books.

Hammer, M., and J. Champy. 1993. *Reengineering the Corporation*. New York: HarperBusiness.

Hannifin, M.J., and C.W. Hughes. 1985. *A Framework for Incorporating Orienting Activities in Computer-Based Interactive Video*. [pub. data n/a].

Harkness, John. 1931. *Calvin: The Man and His Ethics*. New York [pub. data n/a].

Harvey, F.D. 1978. Greeks and Romans Learn to Write. In *Communications Arts in the Ancient World*. Ed. E. Havelock and J. Hershbell. New York: Hasting House.

Havelock, Eric. 1963. *Preface to Plato*. Cambridge, Mass.: Howard Univ. Press.

________. 1976. Origin of Western Literacy. Toronto: Ont. Inst. of Studies in Education.

________. 1978. The Alphabetization of Homer. In *Communications Arts in the Ancient World*. Ed. E. Havelock and J. Hershbell. New York: Hasting House.

Hawlisher, G. 1987. Word Processing, Revision, and College Freshmen. Unpublished ms. cited by Thomas Barker in *Studies in Word Processing and Writing*. [See under Barker].

Her Majesty's Inspectors of Schools. 1984. The Content of the Primary Curriculum: Science. In *The Teaching of Primary Science: Policy and Practice*. Ed. C. Richards and D. Holford. London [pub. data n/a].

Hershbell, J. 1978. The Ancient Telegraph: War and Literacy. In *Communications Arts in the Ancient World*. Ed. E. Havelock and J. Hershbell. New York: Hasting House.

Hertzler, Joyce. 1965. *A Sociology of Language*. New York: Cooper Sq. Pub.

Hickerson, Nancy. 1980. *Linguistic Anthropology*. New York: Holt, Rinehart & Winston.

Hockett, Charles. 1960. The Origin of Speech. *Scientific American* 203{3}.

The Hulton Press Survey. 1948. London: Hulton Press.

Hunter, B. 1983. *My Students Use Computers*. Reston, Va.: Reston Pub. Co.

Innis, Harold. 1951. *The Bias of Communication*. Toronto: Univ. of Toronto Press.

__________. 1972. *Empire and Communications*. With foreword by Marshall McLuhan. Originally pub. by Oxford Univ. Press [1950]. Toronto: Univ. of Toronto Press.

Irwin, M.E. 1987. Connections: Young Children, Reading, Writing, and Computers. *Computers in the Schools* 4{1}.

Iserloh, E. 1968. *The Theses Were Not Posted*. Boston: Beacon Press.

Jackson, B., and O. Marsden. 1962. *Education and the Working Class*. Penguin.

Kenyon, Frederick. 1937. *Books and Readers in Ancient Greece and Rome*. Oxford: Clarendon Press.

King, R., and J. Raynor. 1981. *The Middle Class*. 2d ed. London: Longman.

Kramer, S.N. 1956. *From the Tablets of Sumer*. Indian Hills, Col.: Falcon Press.

_________. 1959. *Life Begins at Sumer*. Garden City, N.Y.: Doubleday Anchor Books.

Kuhn, T.S. 1972. *The Structure of Scientific Revolutions*. Chicago: Univ. of Chicago Press.

Kurland, D.M., and L.C. Kurland. 1987. Computer Applications in Education: A Historical Overview. *Annual Reviews of Computer Science* 2:319-20.

Labbett, Beverley. 1988. Skillful Neglect. In *Breaking into the Curriculum*. Ed. J. Schostak. London: Methuen.

Lepper, M.R., and R.W. Chabay. 1985. Intrinsic Motivation and Instruction: Conflicting Views on the Role of Motivational Processes in Computer-Based Education. *Educational Psychologist* 20{4}.

Levin, J.A., M. Riel, N. Miyake, and M. Cohen. 1987. Education on the Electronic Frontier: Teleapprentices in Globally Distributed Educational Context. *Contemporary Educational Psychology* 12.

Levi-Strauss, C. See Carbonnier, G., *Conversations with Claude Levi-Strauss.*

Lewis, R., and A. Maude. 1949. *The English Middle Class.* London: Phoenix House.

Lieberman, Philip. 1975. *On the Origins of Language.* New York: Macmillan.

Litchfield, B., and S. Mattson. 1989. The Interactive Media Science Project. *Journal of Computers in Mathematics and Science Teaching* 9{1}.

Logan, Robert K. 1979. The Mystery of the Discovery of Zero. *Etcetera.* Vol. 36.

________. 1982. Five Technological Breakthroughs. Cited at Telidon in the Total Information-Communications Environment Workshop, 1-2 March, in Proceedings of the DOC Workshop on Social Psychological Issues in Videotext Development. Univ. of Toronto.

________. 1986a. *The Alphabet Effect.* New York: William Morrow.

________. 1986b. The Use of Computers in Primary and Junior Science Education: A Guide for Classroom Teachers, Consultants and Software Developers. Ont. Min. of Ed. Report. [Dec.]

Logan, Robert, S. Padro, L. Miller, R. Ragsdale, E. Sullivan, J. Brine, E.M. Johnston, B. McKelvey, and H. Wideman. 1983. Research Questions on the Impact of Computers in the Classroom. Ont. Min. of Education. Ed. and Tech. Ser. ONO 4268.

Ludwig, O. 1983. Writing Systems and Written Language. In *Focus.* Ed. F. Coulmas and K. Erlich. New York: Mouton.

Lusseyran, Jacques. 1963. *And There Was Light.* Boston: Little Brown.

Maddux, C.D. 1988. Preface to Assessing the Impact of Computer-Based Instruction. *Computers in the Schools.* Vol. 5.

Malone, T.W. 1981. Toward a Theory of Intrinsically Motivating Instruction. *Cognitive Science* 4.

Mandinach, E. 1989. Model Building and Use of Computer Simulation of Dynamic Systems. *Journal of Educational Computing Research* 5{2}.

Manzelli, J. 1986. New Curriculum Soundings on a Voyage of the Mimi. *Computers in the Schools* 3{1}.

Marchand, Phillip. 1989. *Marshall McLuhan: The Medium and the Messenger.* Toronto: Random House.

Marshack, A. 1964. *Science*. Vol. 146.

Marx, Karl. 1954. *Das Kapital*. Moscow: Progress Publishers.

Marx, Karl and F. Engels. 1940. *The Communist Manifesto*. Toronto: Progress Books.

McLuhan, Marshall. 1962. *The Gutenberg Galaxy: The Making of Typographic Man*. Toronto: Univ. of Toronto Press.

________. 1964. *Understanding Media*. New York: McGraw Hill.

________. 1967. *The Medium Is the Massage*. New York: Bantam Books.

________. 1972. Foreword to *Empire and Communications*. See Innis, Harold.

________. 1977. Letter to Kaj Spenser, Nov. 8, 1977. Cited in *Marshall McLuhan: The Medium and the Messenger* (see Marchand, Phillip).

________. 1988. Media and the Inflation Crowd. *The Antigonish Review*. Nos. 74 & 75 (summer & autumn).

McLuhan, Marshall, and R.K. Logan. 1977. Alphabet, Mother of Invention. *Etcetera*. Vol. 34. [Dec.].

McLuhan, Marshall, and Eric McLuhan. 1988. *Laws of the Media: The New Science*. Toronto: Univ. of Toronto Press.

McLuhan, Marshall, and Barrington Nevitt. 1972. *Take Today: Executives as Dropout*. Don Mills: Longman.

Mehan, H. 1989. Microcomputers in Classrooms: Educational Technology or Social Practice. *Education and Anthropology Quarterly* 20{1}.

Mersich, D. 1982. *Canadian Datasystems*. Vol. 14, No.10. [Oct.].

Meyers, E. 1960. *Education in the Perspective of History*. New York: Harper.

Mill, James. 1969. *James Mill on Education*. London: Cambridge Univ. Press.

Montague, Marjorie. 1990. Computers and Writing Process Instruction. *Computers in the Schools* 7{3}.

Moore, M. 1987. The Effect of Word Processing Technology in a Developmental Writing Program on Writing Quality, Attitudes Towards Composing, and Revision Strategies of 4th and 5th Grade Students. Ph.D. diss., Univ. of South Florida. *Dissertation Abstracts International* 48, 635A.

Moravcsik, Julius. 1967. Aristotle's Theory of Categories. In *Aristotle: A Collection of Critical Essays*. Ed. J. Moravcsik. Garden City, N.Y.: Anchor Books.

Morocco, C., and S. Newman. [date n/a]. Word Processors and the Acquisition of Writing Strategies. *Journal of Learning Disabilities* 19.

Naisbitt, John, and Patricia Aburdene. 1985. *Reinventing the Corporation*. New York: Warner Books.

Olds, H.F. 1986. Information Management: A New Tool for a New Curriculum. *Computers in the Schools* 3{1}.

Olson, John. 1988. *Schoolworlds/Microworlds*. Oxford: Oxford Univ. Press.

Ong, Walter. 1977. *Interfaces of the Word*. Ithaca and London: Cornell Univ. Press.

________. 1982. Orality and Literacy: *The Technologizing of the Word*. London and New York: Methuen.

Osborne, David, and Ted Gaebler. 1992. *Reinventing Government*. Reading, Mass.: Addison-Wesley.

Paivio, A., and I. Begg. 1981. *Psychology of Language*. Englewood Cliffs, N.J.: Prentice-Hall.

Palm, F.C. 1936. *The Middle Class Then and Now*. New York: Macmillan.

Papert, S. 1980. *Mindstorms: Children, Computers and Powerful Ideas*. New York: Basic Books.

Pea, R., and R.D. Kurland. 1986. Cognitive Technologies for Writing. *Review of Research in Education* 13.

Perkin, H. 1969. *The Origin of Modern English Society*. London: Routledge and K. Paul.

Piaget, Jean. 1950. *Introduction à l'Epistomologie Génétique*. Paris: Presses Universitaires de France.

Pitschka, R. 1984. Little CAI in Hollister. In *The Intelligent Schoolhouse*. Ed. D. Peterson. Reston, Va.: Reston Pub. Co.

Pfeiffer, John. 1977. *The Emergence of Society*. New York: McGraw Hill.

Polin, Linda. 1990a. Young Children, Literacy and Computers. *The Computing Teacher* 18{3}.

________. 1990b. Computer Environments and the Organization of Classroom Tasks. *The Computing Teacher* 18{2}.

________. 1990c. Computers for Student Writing: The Relationship Between Writer, Machine, and Text. *The Computing Teacher* 18{7}.

________. 1991. Current Thinking about Critical Thinking: Implications for Technology Use. *The Computing Teacher* 18{5}.

Porter, M., and V. Millar. 1991. How Information Gives You Competitive Advantage. In *Revolution in Real Time: Managing IT in the 1990s. Harvard Business Review*.

Poulantzas, N. 1978. *Classes in Contemporary Capitalism*. London: Verso.

Power, E. 1970. *Main Currents in the History of Education*. 2d. ed. New York: McGraw Hill.

Prigogene, Iya. 1984. *Order Out of Chaos*. New York: Bantam Books.

Putnam, George. 1962. *Books and Their Makers during the Middle Ages*. New York: Hillary House.

Reznick, R., and D. Taylor. 1994. *The Internet Business*. Indianapolis [pub. data n/a.]

Riel, M. 1991–92. Approaching the Study of Networks. *The Computing Teacher*. [Dec./Jan.]

Roberts, K. et al. 1977. *The Fragmentary Class Structure*. London: Heinemann.

Rohman, G. 1965. Pre-writing: The Stage of Discovery in the Writing Process. *College Composition and Communication* 16.

Rondo, Cameron. 1993. *A Concise History of the World*. New York and Oxford: Oxford Univ. Press.

Sahlins, M. 1968. Notes on the Original Affluent Society. In *Man, the Hunter*. Ed. R. Lee and I. Devore. Chicago: Aldine Pub. Co.

Sapir, Edward. 1921. *Language*. New York: Harcourt, Brace.

de Saussure, F. 1967. *Grundfragen der allgemeinen Sprachwissenschaft*. [Basic Questions of Linguistics] 2d ed. Berlin.

Schickel, Richard. 1965. Marshall McLuhan: Canada's Intellectual Comet. Schickel quoting McLuhan in *Harper's* [Nov.].

Schmandt-Besserat, D. 1978. The Earliest Precursor of Writing. *Scientific American* 238.

________. 1979. An Archaic Recording System in the Uruk-Jemdet Nasr Period. *American Journal of Archeology* 83.

________. 1980. The Envelopes That Bear the First Writing. *Technology and Culture* 21, No.3.

________. 1981. From Tokens to Tablets: A Re-Evaluation of the So-Called Numerical Tablets. *Visible Language* 15.

________. 1984a. Before Numerals. *Visible Language* 18.

________. 1984b. The Use of Clay before Pottery in the Zagros. *Expedition* 16, No. 2.

________. 1985. Clay Symbols for Data Storage in the VIII Millennium B.C. In *Studi di Palentologia in onore di Salvatore M. Puglisi*. La Spaienza: Universita di Roma.

________. 1986a. An Ancient Token System: The Precursors to Numerals and Writing. *Archeology*, Vol. 39, No. 6.

________. 1986b. The Origins of Writing. *Written Communication* 13.

________. 1986c. Tokens at Susa. *Oriens Antiquus*, Vol. 25, Nos. 1–2.

________. 1987. Oneness, Twoness, Threeness. *The Sciences*. New York Academy of Sciences.

________. 1988a. Quantification and Social Structure. In *Communication and Social Structure*. Ed. D.R. Maines and C.J. Couch. Springfield, Ill.: C.C. Thomas.

________. 1988b. Tokens at Uruk. *Bagdader Mitteilungen*, Vol. 19.

________. 1992. Before Writing: Vol. 1. *From Counting to Cuneiform*. Houston: Univ. of Texas Press.

Schostak, John. 1988a. Impact on the Curriculum. In *Breaking into the Curriculum*. Ed J. Schostak. London: Methuen.

________. 1988b. Intelligence Communities. In *Breaking into the Curriculum*. Ed. J. Schostak. London: Methuen.

________. 1988c. Introduction. In *Breaking into the Curriculum*. Ed. J. Schostak. London: Methuen.

Schwartz, E. 1991. At This High Tech School "Gum" Is a Requirement. *Business Week*. [Nov. 11].

Schwartz, Joel. 1989. Implementing Computer Literacy in Our Schools: What We Have Learned. *Computers in the Schools* 6{1/2}.

Science Council of Canada. 1984. Science for Every Student: Educating Canadians for Tomorrow's World. Report 36. Ottawa.

Senge, Peter. 1993. *The Fifth Discipline*. New York: Doubleday.

Shibles, Warren A. 1971. *Metaphor, An Annotated Bibliography and History*. Whitewater, Wis.: Language Press.

Sloan, D. 1985. Introduction. In *The Computer in Education*. Ed. D. Sloan. New York: Teachers College Press.

Smith, F. 1982. *Writing and the Writer*. New York: Holt, Rinehart & Winston.

Smith, Ken. 1991. Hypermedia in Education: Implications for Cognition and Curriculum. An unpublished paper submitted as a term paper to the author in a course on media and computers, given in the Dept. of Measurement, Evaluation and Computer Applications at the Ont. Inst. for Studies in Educ., Toronto.

Smith, W. 1955. *Ancient Education*. New York: Philosophical Library.

Speises, Prof. (see Meyers, E.)

Stebbins, B. 1990. HyperCard: The Tool for the Classrooms of Tomorrow. *Computers in the Schools* 7{4}.

Steinberg, S.H. 1955. *Five Hundred Years of Printing*. Middlesex: Penguin Books.

Strickland, A.W., and T. Hoffer. 1989. Databases, Problem Solving and Laboratory Experience. *Journal of Computers in Mathematics and Science Technology* 9{1}.

________. 1990–91. Integrating Computer Databases with Laboratory Problems. *The Computing Teacher*, Dec.–Jan.

Stubbs, Michael. 1980. *Language and Literacy*. London: Routledge & K. Paul.

________. 1982. Written Language and Society. In *What Writers Know*. Ed. M. Nystrand. New York: Academic Press.

Sullivan, E., P. Olson, N. Watson, D. Bors, H. Wideman, J. Zimmerman, and R.K. Logan. 1986. The Development of Policy and Research Projections for Computers in Education: A Comparative Ethnography. Project No. 499-83-0017. Study funded by Social Science and Humanities Research Council of Canada and carried out at Ont. Inst. for Studies in Educ., Toronto.

Tapscott, D., and A. Caston. 1993. *Paradigm Shift*. New York: McGraw Hill.

Taylor, R.P., ed. 1980. *The Computer in the School: Tutor, Tool, Tutee*. New York: Columbia Univ. Teachers College Press.

Teale, W., and E. Sulzby, eds. 1986. *Emergent Literacy: Writing and Reading*. Norwood, N.J.: Ablex Publishing Corp.

Tierney, R. 1989. Student Thinking Process. ACOT Report 3. Apple Computer Inc. ERIC document No. 316-201.

Toffler, Alvin. 1980. *The Third Wave*. New York: William Morrow.

________. 1990. *Powershift*. New York: Bantam Books.

Vygotsky, L.S. 1962. *Thought and Language*. Cambridge, Mass.: M.I.T. Press.

Weber, M. 1946. *Essays in Sociology*. Trans. H. Gerth and C. Wright Mills. Oxford: Oxford Univ. Press.

________. 1948. *The Protestant Ethic and the Spirit of Capitalism*. Trans. T. Parsons. London: Allen and Unwin.

Whorf, Benjamin Lee. 1964. *Language, Thought and Reality: Selected Writings*. Cambridge, Mass.: M.I.T. Press.

Wilson, Peter. 1988. *The Domestication of the Human Species*. New Haven: Yale Univ. Press.

Young, M. and P. Willmot. 1973. London: Routledge and K. Paul.

INDEX